AF707323

Mathematical Modeling for Flow and Transport Through Porous Media

Edited by

GEDEON DAGAN
Tel Aviv, Israel

ULRICH HORNUNG
Neubiberg, Germany

and

PETER KNABNER
Augsburg, Germany

Reprinted from *Transport in Porous Media*
Volume 6, Nos. 5 & 6 (1991)

KLUWER ACADEMIC PUBLISHERS
DORDRECHT / BOSTON / LONDON

Library of Congress Cataloging-in-Publication Data

Mathematical modeling for flow and transport through porous media / edited by Gedeon Dagan, Ulrich Hornung, and Peter Knabner.
p. cm.
"Reprinted from Transport in porous media, volume 6, nos. 5 & 6 (1991)."
Proceedings of an international workshop held at the Schwabisches Bildungszentrum in Irsee, Bavaria on May 15-20, 1989.
Includes index.
ISBN 0-7923-1616-9 (acid free paper)
1. Groundwater flow--Mathematical models--Congresses.
2. Porosity--Mathematical models--Congresses. I. Dagan, G. (Gedeon), 1932- . II. Hornung, Ulrich, 1941- . III. Knabner, Peter.
TC176.M353 1991
551.49'01'5118--dc20 91-46824

ISBN 0-7923-1616-9

Published by Kluwer Academic Publishers,
P.O. Box 17, 3300 AA Dordrecht, The Netherlands.

Kluwer Academic Publishers incorporates
the publishing programmes of
D. Reidel, Martinus Nijhoff, Dr W. Junk and MTP Press.

Sold and distributed in the U.S.A. and Canada
by Kluwer Academic Publishers,
101 Philip Drive, Norwell, MA 02061, U.S.A.

In all other countries, sold and distributed
by Kluwer Academic Publishers Group,
P.O. Box 322, 3300 AH Dordrecht, The Netherlands.

Printed on acid-free paper

All Rights Reserved
© 1991 Kluwer Academic Publishers
No part of the material protected by this copyright notice may be reproduced or utilized in any form or by any means, electronic or mechanical, including photocopying, recording or by any information storage and retrieval system, without written permission from the copyright owner.

Printed in the Netherlands

Table of Contents

Transport in Porous Media **6**: 473, 1991
© 1991 *Kluwer Academic Publishers. Printed in the Netherlands.*

International Workshop on Mathematical Modeling for Flow and Transport Through Porous Media

Irsee, Bavaria

May 15-20, 1989

In the "Schwäbisches Bildungszentrum" Irsee, Bavaria, an international workshop took place on "Mathematical Modeling for Flow and Transport Through Porous Media" during the week May 15-20, 1989. The major topics of the meeting were *free and moving boundary problems, chemical and biological processes, structured media, multiphase flow, scale problems, stochastic aspects,* and *optimization problems.* Points of special interest were also hysteresis and parameter identification. The 57 participants came from 14 different countries.

The conference got very positive resonance from all sides. The success of the workshop had two origins: the conference site is a wonderful place in the Allgäu that is somehow isolated and offers all facilities to support a relaxed atmosphere for discussions and the exchange of ideas. On the other hand, the group of people that came together formed a mixture of scientists from a large variety of disciplines with all kinds of backgrounds. Among them were hydrologists, soil physicists, civil and chemical engineers, environmental scientists, and mathematicians. The basis of common interest was that all of them were experts in a field dealing with *porous media* and their theory. The schedule was kept such that besides invited lectures and contributed papers there was enough time for detailed discussions which were all very stimulating. Different aspects from the various fields of applications but also fundamental questions about modeling and mathematical problems came up. At the end of the week several participants remarked that they had not only met people for the first time who are doing things being very close their own work but also they had learnt about scientific methods and approaches they had not heard about before.

The editors of this volume thank the sponsors (especially the Deutsche Forschungsgemeinschaft and Südchemie, München) for their financial support of the meeting, the staff of the Schwäbisches Bildungszentrum for their friendly service, and all the participants for their papers presented, for the discussions and also for the contributions being published in the proceedings.

Gedeon Dagan (Tel Aviv), Ulrich Hornung (Neubiberg), Peter Knabner (Augsburg)

Transport in Porous Media **6**: 475–477, 1991
© 1991 *Kluwer Academic Publishers. Printed in the Netherlands.*

International Workshop on Mathematical Modeling for Flow and Transport Through Porous Media

Program

May 15-19, 1989

1 FREE AND MOVING BOUNDARY PROBLEMS

Monday a.m.
Chair: K.-H. HOFFMANN (Augsburg)

- 9.00-10.00 **J. C. BRUCH** (St. Barbara) *Fixed domain methods*
- 10.00-10.15 D. KIRKHAM (Ames) *The stream function and generation of boundary conditions in groundwater flow*
- 10.15-10.30 Discussion
- 11.00-11.45 **C. J. VAN DUIJN** (Delft) *Salt and fresh water problems*
- 11.45-12.00 J. KNIGHT (Canberra) *The seepage exclusion problem for airfilled cavities in unsaturated flow*
- 12.00-12.30 Discussion

2 CHEMICAL AND BIOLOGICAL PROCESSES

Monday p.m.
Chair: P. KNABNER (Augsburg)

- 18.30-19.15 **M. F. WHEELER** (Houston) *Modeling of insitu biorestoration in porous media*
- 19.15-19.30 D. LOMEN (Tucson) *A perturbation solution of the equation for transport of a single reactive chemical with non-linear rate loss*
- 19.30-19.45 D. TONDEUR (Nancy) *The kinematic wave approach applied to packed-bed heat and mass flow*
- 19.45-20.00 Discussion
- 20.30-21.15 **S. VAN DER ZEE** (Wageningen) *Reaction kinetics in soil*
- 21.15-21.30 P. KOWALIK (Gdansk) *Modeling plant water uptake and transpiration*
- 21.30-21.45 M. HAUHS (Göttingen) *Models for complex systems*
- 21.45-22.00 Discussion

3 STRUCTURED MEDIA

Tuesday a.m. and p.m.
Chair: BRUINING (Delft)

- 9.00-9.45 **J. DOUGLAS** (West Lafayette) *Simulation of immiscible displacement in fractured reservoirs*
- 9.45-10.00 A. BOURGEAT (St. Etienne) *Critical size of a fissure in a porous medium*
- 10.00-10.30 Discussion
- 11.00-12.00 **S. NEUMAN** (Tucson) *A quasilinear theory of non-Fickian and Fickian dispersion in geologic media*
- 12.00-12.30 Discussion

Chair: U. HORNUNG (Neubiberg)

- 15.00-16.00 **R. SHOWALTER** (Austin) *Diffusion models with microstructure*
- 16.00-16.30 Discussion
- 17.00-17.45 **W. JÄGER** (Heidelberg) *Homogenization and models for a chromatograph*
- 17.45-18.00 M. MENENTI (Wageningen) *Heat and water flow processes in saline structured desert soils*
- 18.00-18.30 Discussion

4 MULTIPHASE FLOW

Wednesday a.m.
Chair: H.-W. ALT (Bonn)

- 9.00-9.45 **R. EWING** (Laramie) *Simulation of multiphase flow*
- 9.45-10.00 P. NICOLAS (Paris) *Liquid/vapour phase change and hysteresis in porous media*
- 10.00-10.30 Discussion
- 11.00-11.45 **Y. MUALEM** (Rehovot) *Hysteresis phenomena*
- 11.45-12.00 G. WEIR (Wellington) *Some outstanding problems in two-phase flow*
- 12.00-12.30 Discussion

5 SCALE PROBLEMS

Thursday a.m. and p.m.
Chair: R. SAMPAIO (Rio de Janeiro)

- 9.00-10.00 **F. A. L. DULLIEN** (Waterloo) *Pore structure*
- 10.00-10.30 Discussion
- 11.00-11.45 **W. G. GRAY** (Notre Dame) *A complete averaging framework for multiphase flow and transport*
- 11.45-12.00 M. A. CELIA, L. Ferrand, W. Soll (Cambridge) *On the use of pore-scale models to derive the pressure-saturation relation for multiphase flow*

- 12.00-12.30 Discussion

Chair: J. KAČUR (Bratislava)

- 18.30-19.15 **S. WHITAKER** (Davis) *Two-phase flow in heterogeneous porous media: a high gradient theory*
- 19.15-19.30 A. BOURGEAT (St. Etienne) *What mathematicians know about Darcy's law*
- 19.30-20.00 Discussion
- 20.30-21.15 **R. A. GREENKORN** (West Lafayette) *Scaling mixing during miscible displacement in heterogeneous porous media*
- 21.15-21.30 J.-F. RODRIGUES (Lisboa) *Asymptotic analysis for the satur-ated-unsaturated flow in multicomponent porous media*
- 21.30-22.00 Discussion

6 STOCHASTIC ASPECTS

Friday a.m.
Chair: R. SMITH (Fort Collins)

- 9.00-10.00 **J. H. CUSHMAN** (West Lafayette) *Stochastic processes in subsurface hydrology*
- 10.00-10.30 Discussion
- 11.00-11.45 **G. DAGAN** (Tel Aviv) *Theory of heterogeneous systems: space versus statistical averaging*
- 11.45-12.00 R. Anlauf, J. RICHTER (Hannover) *Modeling solute transport in heterogeneous soils using Monte-Carlo simulations*
- 12.00-12.30 Discussion

7 OPTIMIZATION PROBLEMS

Friday p.m.
Chair: M. NIEZGODKA (Warsaw)

- 15.00-15.30 P. DUCHATEAU (Fort Collins) *Trace type functional equations and inverse problems*
- 15.30-15.45 W. DURNER (Bayreuth) *The dependence of prediction of the hydraulic conductivity in porous media on the description of the water retention characteristic - a sensitivity analysis*
- 15.45-16.00 D. ZACHMANN, I. White (Fort Collins) *Parameter identification in a linearized flow equation using sorptivity*
- 16.00-16.30 Discussion
- 17.00-17.30 B. HERRLING (Karlsruhe) *Theory and application of optimization methods for problems of groundwater management*
- 17.30-17.45 R. FEDDES, P. Kabat, J. J. B. Bronswijk (Wageningen) *Sinks and sources in modeling transport processes through porous media*
- 17.45-18.00 B. H. GILDING (Twente) *Ricards' equation*
- 18.00-18.30 Discussion

Transport in Porous Media **6**: 479–499, 1991
© 1991 *Kluwer Academic Publishers. Printed in the Netherlands.*

SIMULATION OF MULTIPHASE FLOWS IN POROUS MEDIA

Richard E. Ewing
Departments of Mathematics,
Chemical Engineering, and
Petroleum Engineering
University of Wyoming
Laramie, Wyoming 82071

Abstract. The ability to numerically simulate single phase and multiphase flow of fluids in porous media is extremely important in developing an understanding of the complex phenomena governing the flow. The flow is complicated by the presence of heterogeneities in the reservoir and by phenomena such as diffusion, dispersion, and viscous fingering. These effects must be modeled by terms in coupled systems of nonlinear partial differential equations which form the basis of the simulator. The simulator must be able to handle both single and multiphase flows and the transition regimes between the two. A discussion of some of the aspects of modeling dispersion and viscous fingering is presented along with directions for future work.

The partial differential equation models are convection-dominated and contain important local effects. An operator-splitting technique is used to address these different effects accurately. Convection is treated by time stepping along the characteristics of the associated pure convection problem, and diffusion is modeled via a Galerkin method for single phase flow and a Petrov-Galerkin technique for multiphase regimes. ELLAM (Eulerian-Lagrangian Localized Adjoint Methods) are discussed to effectively treat the advection-dominated processes. Accurate approximations of the fluid velocities needed in the Eulerian-Lagrangian time-stepping procedure are obtained by mixed finite element methods. Adaptive local grid refinement techniques are then indicated to resolve important local phenomena around wells and large heterogeneities or to resolve the moving internal boundary layers which often govern the mass transfer between phases.

Key words. Multiphase flow, numerical simulation, viscous fingering.

1. Introduction

Simulation of complex fluid flow processes in heterogeneous porous media is an important aspect of petroleum reservoir engineering and the modeling and control of contaminant transport through porous media. The purpose of this paper is to address techniques for treating the difficulties caused in simulation by heterogeneities in the reservoir, diffusion and dispersion, viscous fingering, and other highly-localized flow properties such as flow around wells. The common difficulty in modeling these various flow regimes is the inability to treat highly varying length scales in large-scale reservoir simulation.

The processes of both single- and multiphase flow involve convection, or physical transport, of the fluids through a heterogeneous porous medium. The equations used to simulate this flow at a macroscopic level are variations of Darcy's law. Darcy's law has been derived [73,74] via a volume averaging of the Navier-Stokes equations, which govern flow through the porous medium at a microscopic or pore-volume level. The length scale for Navier-Stokes flow (10^{-4}-10^{-3} meters) is very different from the scale required by normal reservoir simulation (10^2-10^3 meters). Reservoirs themselves have scales of heterogeneity ranging from pore-level to field scale. In the standard averaging process for Darcy's law, many important physical phenomena which may eventually govern the macroscopic flow are lost. The further averaging of reservoir and fluid properties necessary to use grid blocks of the size of 10-10^2 meters in field-scale simulators further complicates the modeling process. We discuss certain techniques to try to address these scaling problems.

Many of the enhanced recovery processes are characterized by the chemical and physical interaction of the fluids. Therefore, diffusion and dispersion are often critical to the flow processes and must be understood and modeled. Molecular diffusion is typically quite small. However, dispersion, or the mechanical mixing caused by velocity variations and flow through heterogeneous rock, can be extremely important and should be incorporated in some way in our models.

Since the mixing and velocity variations are influenced at all relevant length scales by the heterogeneous properties of the reservoir, much work must be done in volume averaging of terms like porosity and permeability. Statistical methods have shown promise in this area [2,38]. Statistical techniques are currently being considered to obtain effective permeability tensors for large-scale models of flow through highly fractured media. However, if the fractures or field-scale heterogeneities are sufficiently large and can be identified, they should be incorporated in the model via special gridding and high permeability variations. If the fractures are sufficiently large, perhaps Darcy's law will not be a good model for flow through the these fractures. The adaptive local grid-refinement techniques presented in this paper can be very valuable in these applications.

Section 2 presents the model equations used in field-scale simulators to describe single and multiphase flow and relates these process models with regard to the diffusion/dispersion concepts. In Section 3, the importance and modeling difficulties of hydrodynamic dispersion and viscous fingering are discussed. Section 4 discusses the use of mixed finite element methods to obtain accurate fluid velocities, while Section 5 presents operator-splitting methods to avoid standard upstream weighting. Section 6 then describes various adaptive grid-refinement techniques which offer promise in resolving important local phenomena in large-scale simulators in an efficient manner.

2. Equations for Single and Multiphase Flow

Although the techniques that we discuss will apply equally well to the recovery of hydrocarbon and the transport of contaminants through the saturated or unsaturated soil zones, we will describe the flow processes in the terminology of miscible and immiscible displacement of resident fluids by injected fluids. The miscible displacement of one incompressible fluid by another, completely miscible with the first, in a horizontal porous reservoir $\Omega \subset \mathbb{R}^2$ over a time period $J = [T_0, T_1]$, is given by

$$-\nabla \cdot \left(\frac{k}{\mu}\nabla p\right) \equiv \nabla \cdot \mathbf{u} = q\ , \quad x \in \Omega\ , \quad t \in J\ , \tag{2.1}$$

$$\phi\frac{\partial c}{\partial t} - \nabla \cdot (\mathbf{D}\nabla c - \mathbf{u}c) = q\tilde{c}\ , \quad x \in \Omega\ , \quad t \in J\ , \tag{2.2}$$

where p and $\mathbf{u}$ are the pressure and Darcy velocity of the fluid mixture, ϕ and k are the porosity and the permeability of the medium, μ is the concentration-dependent viscosity of the mixture, c is the concentration of the invading fluid, q is the external rate of flow, and $\tilde{c}$ is the inlet or outlet concentration. The form of the diffusion-dispersion tensor $\mathbf{D}$ is given by

$$\mathbf{D} = \phi(x)\left\{d_m I + |\mathbf{u}|\left(d_l E(\mathbf{u}) + d_t E^{\perp}(\mathbf{u})\right)\right\} \tag{2.3}$$

where

$$E_{ij}(\mathbf{u}) = \frac{\mathbf{u}_i \mathbf{u}_j}{|\mathbf{u}|^2}\ , \tag{2.4}$$

$E^{\perp} = I - E$, d_m is the molecular diffusion coefficient, and d_l and d_t are the longitudinal and transverse dispersion coefficients, respectively. In general, $d_l \approx 10dt$, but this may vary greatly with different soils, fractured media, etc. The viscosity μ in Equation (2.1) is assumed to be determined by some mixing rule. In addition to Equations (2.1) and (2.2), initial and no flow boundary conditions are specified. The flow at injection and production wells is modeled in Equations (2.1) and (2.2) via point sources and sinks.

The equations describing two phase, immiscible, incompressible displacement in a horizontal porous medium are given by

$$\phi\frac{\partial S_w}{\partial t} - \nabla \cdot \left(k\frac{k_{rw}}{\mu_w}\nabla p_w\right) = q_w\ , \quad x \in \Omega\ , \quad t \in J\ , \tag{2.5}$$

$$\phi\frac{\partial S_o}{\partial t} - \nabla \cdot \left(k\frac{k_{ro}}{\mu_o}\nabla p_o\right) = q_o\ , \quad x \in \Omega\ , \quad t \in J\ , \tag{2.6}$$

where the subscripts w and o refer to water and oil respectively, S_i is the saturation, p_i is the pressure, k_{ri} is the relative permeability, μ_i is the viscosity, and q_i is the external flow rate, each with respect to the ith phase. The saturations sum to unity.

One of the saturations can be eliminated; let $S = S_w = 1 - S_o$. The pressure between the two phases is described by the capillary pressure

$$p_c(S) = p_o - p_w\ . \tag{2.7}$$

Although formally, the equations presented in (2.1) and (2.2) seem quite different from those in (2.5) and (2.6), the latter system may be rearranged in a form which very closely resembles the former system. In order to use the same basic simulator in our sample computations to treat both miscible and immiscible displacement, we will follow the ideas of Chavent [11], and utilize a miscible/immiscible flow analogy.

The following equations are obtained:

$$-\nabla \cdot (k\lambda(S)\nabla p) = q_w + q_o = q_t \ , \tag{2.8}$$

$$v_t = -k\lambda(S)\nabla p \ . \tag{2.9}$$

$$\phi\frac{\partial S}{\partial t} + \nabla \cdot \left(k\lambda(S)\overline{\lambda}_o\frac{dp_c}{dS}\nabla S\right) + \nabla \cdot \left(\overline{\lambda}_w v_t\right) = q_w \ . \tag{2.10}$$

Here p and v_t denote the total fluid pressure and velocity, respectively, and $\lambda(S)$ and $\overline{\lambda}_i(S)$ are the total and fractional phase mobilities, respectively (see [21]).

The equations presented above describe miscible and immiscible flow in porous media. They can be used to simulate various production strategies in an attempt to understand and hopefully optimize hydrocarbon recovery. However, in order to use these equations effectively, parameters that describe the rock and fluid properties for the particular reservoir application must be input. The relative permeabilities which are nonlinear functions of water saturation can be estimated via laboratory experiments using reservoir cores and resident fluids [10]. Similarly, fluid viscosities are relatively easy to obtain. However, the permeability tensor k, the porosity ϕ, the capillary pressure curve $p_c(S)$, and the diffusion and dispersion coefficients are effective values that must be obtained from local properties via scaling techniques. In addition, the inaccessibility of the reservoir to measurement of even the local properties increases the difficulties. See [20,27,48] and the references contained therein for a survey of parameter estimation and history-matching techniques which have been applied. Section 3 below then describes the modification of estimated viscosity, permeability, or dispersion coefficients to incorporate the important effects of viscous fingering. Much work is needed in this research area.

The accurate modeling of permeabilities, which may vary considerably from grid block to grid block, causes more difficulties in reservoir simulation. For flow parallel to permeability discontinuities, an arithmetic averaging of permeabilities is appropriate. However, for flow across discontinuities, a harmonic averaging of permeabilities is required. With high velocity variations, this averaging prescription is not easy to maintain. The mixed finite element methods presented in [16,17,28-30,32,33,67] insure these averaging rules and the simultaneous approximation of the Darcy velocity directly. These techniques have been extremely successful even with fairly large grid blocks and in difficult computational simulations in industrial applications [28,29]. Their combination with local grid refinement around transmissibility changes holds even more potential.

3. Dispersion and Viscous Fingering

The effects of dispersion in various flow processes have been discussed extensively in the literature [46,57,58,75,76,80]. Russell and Wheeler [67] and Young [83] have given excellent surveys of the influence of dispersion and attempts to incorporate it in present reservoir simulators. Various terms which affect the length of the dispersive mixing zone include viscosity and velocity variations and reservoir heterogeneity. Much work is needed to quantify these effects and to obtain useful effective dispersion coefficients for field-scale simulators. The dispersion tensor (see Equation (2.3)) has strong velocity dependence [21,57,67]. The longitudinal dispersion is often an order of magnitude larger than the transverse dispersion. This variation enhances unstable flow regimes induced by viscosity differences and reservoir heterogeneity. Initial work on correlation of dispersion coefficients presented with statistical simulations in [34] will be summarized below.

In both miscible and immiscible displacements, the process of pushing a heavy, viscous oil through a heterogeneous porous medium with a lighter, less viscous fluid can be a very unstable one. If the flow rate is sufficiently high, the interface between the resident petroleum and the invading fluid becomes unstable and tends to form long fingers which grow in length toward the production wells, bypassing much of the hydrocarbon. Once a path consisting of the injected fluid has extended from an injection well to a production well, that production well will henceforth produce primarily the injected fluid, which flows more easily due to its lower viscosity and higher mobility. The production of petroleum from that well is then greatly reduced. This phenomenon, termed *viscous fingering*, is well known [4,12,39,40,44,47,51-64,68-72,78,79,81,82] and is a serious problem in hydrocarbon recovery. This problem and different techniques for understanding and modeling it were surveyed in [26].

The equations used to describe fluid flow in porous media are obtained by applying various averaging techniques to the Navier-Stokes equations which describe the macroscopic flow properties. The averaging is done over tubes of varying geometry and tortuosity. Clearly, the Navier-Stokes equations do not describe the flow on a macroscopic level. Analysis must be applied to the Navier-Stokes equations to understand the conditions under which small flow perturbations caused by tortuous flow in the porous media will become unstable and will grow into large fingers which affect the flow on a macroscopic level. The understanding of this instability is crucial, both in attempts to stabilize the flow via polymers, etc., and to model the fingering phenomenon when it cannot be controlled. There are three distinct problems associated with the viscous fingering phenomenon. Knowledge of the conditions causing the onset of the instability is one major goal. Then the understanding of the nonlinear effects which cause certain fingers to grow preferentially and coalesce to form larger fingers is essential. The rate of this finger growth must also be determined. Once the growth of the fingers is understood, from a microscopic to a macroscopic level, statistical methods should be used to incorporate the large-scale effects of fingering into the mathematical models used to simulate the field scale process without trying to treat the individual fingers.

An important influence upon the finger growth is the velocity dependence of the dispersion tensor. Once a finger is initiated, flow is preferentially directed along the finger axis. Then, since the longitudinal dispersion (in the direction of flow) is considerably larger than the transverse dispersion, the dispersive effects tend to increase finger growth. Therefore, a complete understanding of the viscous fingering phenomena is not possible without a better model for the dispersion tensor and the relative length scales of the mixing zone and the finger size.

When viscous fingering occurs, the finger wavelengths are generally much smaller than the typical computational cells which are used in a simulator, and therefore cannot be represented on a standard numerical grid. In order to correctly model the physics of the fingering phenomenon, one would have to add a prohibitive number of grid points around the moving front. If the fingering effect is not modeled, in general an overly optimistic value of fluid recovery is predicted. Realizing the inability to realistically model the physics of fingering with reasonable grids, a lumped-parameter approach, similar in principle to the averaging of the Navier-Stokes flow equations which was previously discussed, has been tried. The original attempt along these lines was due to Koval [45], utilizing a global Buckley-Leverett approach. This approach was extended by Todd and Longstaff [77] via introduction of a mixing parameter w.

Other attempts to incorporate the effects of fingering in a simulator without trying to model the physics on its smaller scale are discussed by Gardner and Ypma [37] and by Fishlock and Rodwell [36]. See [7] for a survey of some of this work together with a method for incorporating velocity variation in the modeling of viscous fingering. None of these attempts is completely satisfactory and the author feels this is an important area for future research. Attempts to use statistics of the flow to empirically determine the longitudinal and transverse dispersion coefficients in the dispersion tensor (see Equation (2.3)) to incorporate fingering effects were presented in [34].

In [34], Ewing *et al.* indicated the ability of coarse-grid dispersion models to match results of both laboratory experiments and fine-grid simulations on highly heterogeneous meshes. The use of a dispersion tensor avoided the optimistic recovery predictions often attributed to standard convection-diffusion models. Heterogeneity, by itself, was shown to be less of an influence on recovery than viscous fingering from adverse mobility ratios. The mixing parameter approach [77] predicted too early breakthrough and optimistic ultimate recoveries compared to experiments and the dispersion models. More research is needed in this area and is currently underway. Studies are currently underway to extend the global effective dispersion concept to multiphase flow in an analogous manner.

Given the importance of dispersion and viscous fingering in the modeling of most enhanced recovery processes, research must progress in several directions. First, the averaging processes used to change length scales must be improved, perhaps via more statistical techniques, to obtain better effective reservoir coefficients for the macroscopic models presented in the next section. Simultaneously, the effective length scales of dispersion and its effect upon viscous fingering must be better understood. Third, better macroscopic techniques for including the effects (perhaps statistical) of viscous fingering and dispersion must be developed. Finally, since the fluid interactions are so critical to enhanced recovery procedures, we cannot afford to continue to use coarse grids and standard upstream-weighting techniques around the fluid interfaces in our simulators, because these methods introduce an artificial diffusion [21], which dominates the physical diffusion/dispersion and smears fronts over coarse grid blocks. This involves higher resolution of the fluid interfaces which, hopefully, can be achieved through adaptive, local grid-refinement techniques presented in this paper.

4. Mixed Methods for Accurate Velocity Approximations

There are two major sources of error in the methods currently being utilized for finite difference discretization of Equations (2.8) and (2.10). The first occurs in the approximation of the fluid pressure and velocity. The second comes from the techniques for upstream weighting to stabilize Equation (2.10). We first describe mixed finite element methods for the accurate approximation of the total velocity $\mathbf{v}_t$. We then discuss some alternate upstream-weighting techniques in a finite element context in the next section for use in Equation (2.10).

Since the diffusion-dispersion term in Equation (2.10) is governed by the fluid velocity, accurate simulation requires an accurate approximation of the velocity $\mathbf{v}_t$. Because the lithology in the reservoir can change abruptly, causing rapid changes in the flow capabilities of the rock, the coefficient k in Equations (2.8) and (2.10) can be discontinuous. In this case, in order for the flow to remain relatively smooth, the pressure changes extremely rapidly. Thus, standard procedures of solving Equation (2.8) as an elliptic partial differential equation for pressure, differentiating or differencing the result to approximate the pressure gradient, and then multiplying by the discontinuous $k\lambda$ can produce very poor approximations to the velocity $\mathbf{v}_t$ [32]. In this section, a mixed finite element method for approximating $\mathbf{v}_t$ and p simultaneously, via a coupled system of first order partial differential equations, will be discussed. This formulation allows the removal of singular terms in the equations and accurately treats the problem of rapidly changing flow properties in the reservoir. In [32] and [28], the benefits of the use of mixed finite elements over standard pressure formulations in a finite element setting were illustrated.

The coupled system of first order equations used to define our methods arises from Darcy's Law and conservation of mass. The system is given by

$$\mathbf{u} = -\frac{k}{\mu}\nabla p\ , \qquad \mathbf{x} \in \Omega\ , \tag{4.1}$$

$$\nabla \cdot \mathbf{u} = q\ , \qquad \mathbf{x} \in \Omega\ , \tag{4.2}$$

subject to a no-flow boundary condition. Clearly, Equations (4.1) and (4.2) will determine p only to within an additive constant. Thus, a normalizing constraint such as $\int_\Omega p(\mathbf{x})d\mathbf{x} = 0$ or $p(\mathbf{x}_s) = 0$, for some $\mathbf{x}_s \in \Omega$, is required in the computation to prevent a singular system.

We next define certain function spaces and notation. Let $W = L^2(\Omega)$ be the set of all functions on Ω whose square is finite integrable. Let $H(\text{div};\Omega)$ be the set of vector functions $\mathbf{v} \in [L^2(\Omega)]^2$ such that $\nabla \cdot \mathbf{v} \in L^2(\Omega)$ and let

$$V = H(\text{div};\Omega) \cap \{\mathbf{v}\cdot\mathbf{n} = 0 \text{ on } \partial\Omega\}\ . \tag{4.3}$$

Let $(v,w) = \int_\Omega vw\ dx$, $\langle v,w\rangle = \int_{\partial\Omega} wv\ ds$, and $||v||^2 = (v,v)$ be the standard L^2 inner products and norm on Ω and $\partial\Omega$. We obtain the weak solution form of Equations (4.1)and (4.2) by dividing each side of Equation (2.1) by k/μ, multiplying by a test function $\mathbf{v} \in V$, and integrating the result to obtain

$$\left(\frac{\mu}{k}\mathbf{u},\mathbf{v}\right) = (p, \nabla \mathbf{v})\ , \qquad \mathbf{v} \in V\ . \tag{4.4}$$

The right-hand side of Equation (4.4) was obtained by further integration by parts and use of the boundary condition. Next, multiplying Equation (4.2) by $w \in W$ and integrating the result, we complete our weak formulation, obtaining

$$(\nabla \cdot \mathbf{u}, w) = (q, w)\ , \qquad w \in W\ . \tag{4.5}$$

For a sequence of mesh parameters $h > 0$, we choose finite dimensional subspaces V_h and W_h with $V_h \subset V$ and $W_h \subset W$ and seek a solution pair $(\mathbf{U}_h; P_h) \in V_h \times W_h$ satisfying

$$\left(\frac{\mu}{k}\mathbf{U}_h,\mathbf{v}_h\right) - (P_h, \operatorname{div}\ \mathbf{v}_h) = 0\ , \qquad \mathbf{v}_h \in V_h\ , \tag{4.6}$$

$$(\operatorname{div}\ \mathbf{U}_h, w_h) = (q, w_h)\ , \qquad w_h \in W_h\ . \tag{4.7}$$

We can now complete the description of our mixed finite element methods with a discussion of particular choices of V_h and W_h (see [30] and [35] for examples).

(4.6)-(4.7) leads to a saddle-point problem requiring care in solution. Preconditioning or efficient iterative methods are essential. Effective block preconditioners are presented in [30,35], and efficient multigrid techniques are discussed in [1].

For problems with smooth coefficients and smooth forcing functions, standard approximation theory results show that, by using higher-order basis functions, correspondingly higher-order convergence rates can be obtained [16,17]. For the fluid flow in porous media applications, the source and sink terms q are sums of Dirac delta functions. As shown by Ewing and Wheeler [35], the resulting smoothness of $\mathbf{v}_t$ is reduced; $\mathbf{v}_t$ is not square ingegrable and thus using the mixed methods, the velocity approximations *would not converge at the wells.* This result was obtained theoretically by Ewing and Wheeler [35] and Douglas, Ewing, and Wheeler [16-17], and computationally by Ewing, Koebbe, *et al.* [30]. By removing the leading term of the singularities, the remaining parts of the velocities are now sufficiently smooth that the approximations to these parts converge at the wells.

We use mixed method techniques for accurately approximating the total velocity $\mathbf{v}_t$. The transport equation (2.10) requires a phase velocity in regimes where two-phase flow occurs. Ewing and Heinemann [29] discussed finite element weightings of the pressure obtained from standard finite difference codes which resulted in significantly better phase velocities than via standard upstream-weighting methods. These finite element weightings are similar to the mixed method ideas presented above, but can be implemented even more easily in existing reservoir simulation codes.

5. Operator-Splitting Techniques

In finite difference simulators, the convection is stabilized via upstream-weighting techniques. In a finite element setting, we use a possible combination of a modified method of characteristics and Petrov-Galerkin techniques to treat the transport separately in an operator-splitting mode.

In miscible or multicomponent flow models, the convective, hyperbolic part is a linear function of the velocity. An operator-splitting technique has been developed to solve the purely hyperbolic part by time stepping along the associated characteristics [18,30,33,65]. We first obtain the non-divergence form of (2.2) by using the product rule for differentiation on the $\nabla \cdot \mathbf{u}c$ term and applying (2.1) to obtain

$$\phi \frac{\partial c}{\partial t} + \mathbf{u} \cdot \nabla c - \nabla \cdot \mathbf{D} \nabla c = q(\tilde{c} - c) . \tag{5.1}$$

Next, the first and second terms in Equation (5.1) are combined to form a directional derivative along what would be the characteristics for the equation if the tensor $\mathbf{D}$ were zero. The resulting equation is

$$\nabla \cdot (\mathbf{D} \nabla c) + q(\tilde{c} - c) = \phi \frac{\partial c}{\partial t} + \mathbf{u} \cdot \nabla c \equiv \phi \frac{\partial c}{\partial \tau} . \tag{5.2}$$

The system obtained by modifying Equations (2.1)-(2.2) in this way is solved sequentially. An approximation for $\mathbf{u}$ is first obtained at time level $t = t^n$ from a solution of Equation (2.1) with the fluid viscosity μ evaluated via some mixing rule at time level t^{n-1}. Equation (2.1) can be solved as an elliptic equation for the pressure, p, or via a mixed finite element method for a more accurate fluid velocity. Let $C^n(x)$ and $\mathbf{U}^n(x)$ denote the approximations of $c(x,t)$ and $\mathbf{u}(x,t)$, respectively, at time level $t = t^n$. The directional derivative is then discretized along the "characteristic" mentioned above as

$$\phi \frac{\partial c}{\partial \tau}(x, t^n) \approx \phi \frac{C^n(c) - C^{n-1}(\overline{x}^{n-1})}{\Delta t} , \tag{5.3}$$

where $\overline{x}^{n-1}$ is defined for an x as

$$\overline{x}^{n-1} = x - \frac{\mathbf{U}^n(x) \Delta t}{\phi} . \tag{5.4}$$

This technique is a discretization back along the "characteristic" generated by the first order derivatives from Equation (5.2). Although the advection-dominance in the original Equation (5.2) makes it non-self-adjoint, the form with directional derivatives is self-adjoint and discretization techniques for self-adjoint equations can be utilized. This modified method of characteristics can be combined with either finite difference or finite element spatial discretizations.

In immiscible or multiphase flow, the convective part is nonlinear. A similar operator-splitting technique to solve this equation needs reduced time steps because the pure hyperbolic part may develop shocks. Recently, an operator-splitting technique has been developed for immiscible flows [13,14,19] which retains the long time steps in the characteristic solution without introducing serious discretization errors.

The operator splitting gives the following set of equations:

$$\phi \frac{\partial \overline{S}}{\partial t} + \frac{d}{dS}\mathbf{f}^m(\overline{S}) \cdot \nabla \overline{S} \equiv \phi \frac{d}{d\tau}\overline{S} = 0 , \tag{5.5}$$

$$\phi \frac{\partial S}{\partial \tau} + \nabla \cdot (\mathbf{b}^m(S)S) - \epsilon \nabla \cdot (D(S)\nabla S) = \mathbf{q}(\mathbf{x}, t) , \tag{5.6}$$

$t_m \leq t \leq t_{m+1}$, together with proper initial and boundary conditions. As noted earlier, the saturation S is coupled to the pressure/velocity equations, which will be solved by mixed finite element methods [16,17,29,30,33].

The splitting of the fractional flow function into two parts: $\mathbf{f}^m(S) + \mathbf{b}(S)S$, is constructed [19] such that $\mathbf{f}^m(S)$ is linear in the shock region, $0 \leq S \leq S_1 < 1$, and $\mathbf{b}(S) \equiv 0$ for $S_1 \leq S \leq 1$. Further, Equation (5.5) produces the same unique physical solution as

$$\frac{\partial S}{\partial t} + \nabla \cdot (\mathbf{f}^m(S) + \mathbf{b}(S)S) = 0 \tag{5.7}$$

with an entropy condition imposed. This means that, for a fully developed shock, the characteristic solution of Equation (5.5) always will produce a unique solution and, as in the miscible case, we may use long time steps Δt without loss of accuracy.

The solution of Equation (5.6) via variational methods leads to the following Petrov-Galerkin equations:

$$B(S_h^m, \phi_i) \equiv (S_h^{m+1}, \phi_i) - \left(\frac{\Delta t}{\phi}\mathbf{b}(\mathbf{x}, t^m) S_h^{m+1}, \nabla \phi_i\right) + \left(\frac{\epsilon \Delta t}{\phi} D(\mathbf{x}, t^m) \nabla S_h^{m+1}, \nabla \phi_i\right)$$

$$= (g_h^m(\mathbf{x}, t^m), \phi_i) , \qquad i = 1, 2, \cdots, N , \qquad S_h^m \in M_h , \qquad \phi_i \in N_h \tag{5.8}$$

where M_h and N_h are the trial and test spaces spanned by $\{\theta_i\}$ and $\{\phi_i\}$, $i = 1, 2, \cdots, N$, respectively. $B(\cdot, \cdot)$ given by Equation (5.8) is an unsymmetrical bilinear form with spatially-dependent coefficients.

In order to obtain Equation (5.8), we have used the characteristic solution from Equation (5.5) to approximate $(\partial/\partial\tau)S$ and the nonlinear coefficients [19] in Equation (5.6). The nonsymmetry in the bilinear form $B(\cdot, \cdot)$ is caused by the nonlinearity of the convective part of the equation, represented by the term $\mathbf{b}(S)S$. This term balances the diffusion forces in the shock region after a traveling front has been established.

We want to use numerical techniques which work well for the symmetric, coercive, bilinear forms to solve Equation (5.8). We consider a procedure, developed by Barrett and Morton [3], which symmetrizes the bilinear form $B(\cdot, \cdot)$ by defining a new set of test functions as follows:

$$B(S^m, \phi_i) = \left(a_{kl} \frac{\partial}{\partial x_k} S^m, \frac{\partial}{\partial x_l} \theta_i\right) \equiv B^*(S^m, \theta_i) , \qquad 0 < a_{kl} < K . \tag{5.9}$$

The test functions ϕ_i defined by Equation (5.9) have nonlocal support and would thus cause serious computational problems for large-scale problems. However, a localization procedure was developed by Demkowitz and Oden [15] which allows efficient computational procedures. Since the bilinear form $B(\cdot,\cdot)$ is coercive, we obtain optimal approximation properties in the norms defined by the form. For computational reasons, it may be better to use an approximate form of the optimal test function ϕ_i. An estimate for the error introduced by an approximate symmetrization of $B(\cdot,\cdot)$ is given by Barrett and Morton [3].

It seems natural to relate the size of the coarse domains to the solution of the pressure-velocity equation [19], since the velocity varies slowly and defines a natural long space scale compared to the variation of the saturation S at a front. A local error estimate which determines if a coarse-grid block must be refined, is given in reference [19]. Normally, local refinement must be performed if a fluid interface is located within the coarse-grid block in order to resolve the solution there. A slightly different strategy is to make the region of local refinement big enough such that we can use the same refinements for several of the large time steps allowed by the method. The local grid-refinement strategy combined with the operator splitting is defined in the literature [13,14,19]. The solution at each of the coarse-grid vertices and the local refinement calculation may be sent to separate processors to achieve a high level of parallelism in the solution process.

The difficult problem with these techniques is the communication of the solution between the fine and coarse grids. The domain decomposition technique described in [5] gives accurate and efficient treatment of the communication problem.

Unfortunately, the modified method of characteristics techniques described above generally do not conserve mass. Also, the proper method for treating boundary conditions in a conservative and accurate manner using these techniques is not obvious. Recently, M.A. Celia, T.F. Russell, I. Herrera, and the author have devised Eulerian-Lagrangian localized adjoint methods (ELLAM) [9,43], a set of schemes that are defined expressly for conservation of mass properties.

The ELLAM formulation was motivated by localized adjoint methods [8,41], which are one form of the optimal test function methods discussed above [3,13,15,19]. We briefly describe these methods. Let

$$Lu = f\ , \qquad x \in \Omega \ \text{ or } \ (x,t) \in \Omega\ , \tag{5.10}$$

denote a partial differential equation in space or space-time. Integrating against a test function ϕ, we obtain the weak form

$$\int_\Omega Lu\phi d\omega = \int_\Omega f\phi d\omega\ . \tag{5.11}$$

If we choose test functions ϕ to satisfy the formal adjoint equation $L^*\phi = 0$ and $\phi = 0$ on the boundary, except at $\partial\Omega$ certain nodes or edges denoted by ℓ_i, then integration by parts (the divergence theorem in higher dimensions) yields

$$\sum_i \int_{\ell_i} uL^*\phi d\omega = \int_\Omega f\phi d\omega\ . \tag{5.12}$$

Various different test functions can be used to focus upon different types of information. Herrera has built an extensive theory around this concept; see [41] for references. The theory is quite general and can deal with situations where distributions do not apply, such as when both u and ϕ are discontinuous.

As in the work of Demkowitz and Oden [15], we want to localize these test functions to maintain sparse matrices. Certain choices of space-time test functions which are useful for linear equations of the form (2.2) have been described in [15,66]. For examples of nonlinear applications of the form (2.10), see [42].

6. Adaptive Grid-Refinement Techniques

Many time-dependent fluid flow problems involve both large-scale processes and highly localized phenomena that are often critical to the overall chemical and physical behaviors of the flows. For large-scale applications, it is frequently impossible to use a uniform grid which is sufficiently fine to resolve the local phenomena without yielding numbers of unknowns that will overburden even the largest of today's supercomputers. Since these local processes are often dynamic, efficient numerical simulation requires the ability to perform dynamic self-adaptive local grid refinement. The need for adaptive techniques has provided the impetus for the development of local grid-refinement software tools, some of which are used in day-to-day applications for small- to mid-size problems. Software and engineering tools capable of dynamic local grid refinement need to be developed for large-scale, fluid flow applications. The adaptive grid-refinement algorithms must also be closely matched with the architecture features of the new advanced computers to take advantage of possible vector and parallel capabilities.

The local patch refinement techniques [5,31,49,50] have proven to be very effective for obtaining local resolution around fixed singular points such as wells in a reservoir. We will discuss the patch approximation technique in the context of local refinement around a point like a well. The major input and output from a reservoir in various production procedures is through wells. Hence, it is important to obtain an accurate approximation to the flow nearby.

We will consider a simple example problem to illustrate our local refinement techniques. We want to approximate the pressure p of the fluid described by either Equation (2.1) or Equation (2.8).

First, we consider the matrix A^c, generated by a finite element or finite difference approximation of the Equations (2.1) or (2.8) using a coarse quasi-uniform mesh. Let the solution P of the original coarse grid problem be decomposed in the form $P = (P_1, P_b, P_2)^T$, where P_1, P_2, and P_b are the parts of the coarse-grid solution in two separate domains Ω_1, Ω_2 and the intersection of the boundary of Ω_1 and Ω_2, respectively. The corresponding decomposition of the matrix A^c can be described in

$$A^c \begin{pmatrix} P_1 \\ P_b \\ P_2 \end{pmatrix} = \begin{pmatrix} A_{11}^c & A_{1b}^c & 0 \\ A_{b1}^c & A_{bb}^c & A_{b2}^c \\ 0 & A_{2b}^c & A_{22}^c \end{pmatrix} \begin{pmatrix} P_1 \\ P_b \\ P_2 \end{pmatrix}. \tag{6.1}$$

We assume that a code exists or can be easily written to solve (6.1) for a quasi-uniform grid which can be highly vectorized to take advantage of the banded structure of the matrix $\overline{A}^c$ which is equivalent to A^c except utilizing a standard lexicographical ordering of the unknowns.

Next, assume that due to some identified localized process, grid refinement is desired in Ω_2. Let P_r be the new approximation on the refined grid in Ω_2 and A_{rr} be the local matrix on Ω_2. Let A_{br} and A_{rb} be the new connection matrices between the interface between Ω_1 and Ω_2 and the refined grid on Ω_2. Then, in order to maintain the sparsity of the composite grid matrix and a simple data structure obtained by concatonating P_r to P, we can write the composite matrix problem in the form

$$\tilde{A}\tilde{P} = \begin{pmatrix} A_{11}^c & A_{1b}^c & 0 & 0 \\ A_{b1}^c & A_{bb}^c & 0 & A_{br} \\ 0 & 0 & I & 0 \\ 0 & A_{rb} & 0 & A_{rr} \end{pmatrix} \begin{pmatrix} P_1 \\ P_b \\ P_2 \\ P_r \end{pmatrix} = \begin{pmatrix} f_1 \\ f_2 \\ 0 \\ f_3 \end{pmatrix}. \tag{6.2}$$

We note that the I on the diagonal of (6.2) and the zeroes in the corresponding row, column, and right-hand side enforce the removal of P_2 from the system without destroying the relationship of

$$\begin{pmatrix} A_{11}^c & A_{1b}^c \\ A_{b1}^c & A_{bb}^c \end{pmatrix} \tag{6.3}$$

to A^c and hence $\overline{A}^c$.

Bramble *et al.* [5] defined a preconditioned conjugate gradient iterative procedure for the efficient solution of the composite problem (6.2). The preconditioner involves two local solutions on the refined grid in Ω_2 and one on the original coarse grid Ω and yields of symmetric preconditioner. Below, we will describe a simpler iterative process which utilizes similar inversions on the refined grid Ω_2 and the coarse grid Ω.

Given a previous iterate for P_b, denoted P_b^n, we solve the local problem on Ω_2 with Dirichlet conditions on the interface between Ω_1 and Ω_2 (given by $A_{rb}P_b$):

$$P_r^n = A_{rr}^{-1}(f_3 - A_{rb}P_b^n). \tag{6.4}$$

This problem can be solved exactly or approximately by some iterative technique. This step could be considered as the first part of a block Gauss-Seidel iterative procedure for the solution of (6.1). The next step would be to use the approximation for P_r^n and then invert the block (6.3) to obtain an approximation for P_1^n and P_b^n. Since this block involves a complex region and may not be well-conditioned, we use an alternate solution method which involves a preconditioner, denoted by B, for the composite matrix $\tilde{A}$.

Using B, we define, for each iterate n and an iteration parameter τ,

$$\tilde{P}^{n+1} = \tilde{P}^n + \tau B^{-1}(\tilde{f} - \tilde{A}\tilde{P}^n). \tag{6.5}$$

Let Q be the residual vector given by

$$\tilde{f} - \tilde{A}\tilde{P}^n = \begin{pmatrix} f_1 - A^c_{11}P^n_1 - A^c_{1b}P^n_b \\ f_2 - A^c_{b1}P^n_1 - A^c_{bb}P^n_b - A_{br}P^n_r \\ 0 \\ f_3 - A_{rb}P^n_b - A_{rr}P^n_r \end{pmatrix} \equiv \begin{pmatrix} Q^n_1 \\ Q^n_2 \\ 0 \\ Q^n_4 \end{pmatrix}. \tag{6.6}$$

Next, we solve the original coarse grid problem

$$A^c \begin{pmatrix} W^{n+1}_1 \\ W^{n+1}_b \\ W^{n+1}_2 \end{pmatrix} = \begin{pmatrix} Q^n_1 \\ Q^n_2 - A_{br}A^{-1}_{rr}Q^n_4 \\ 0 \end{pmatrix} \tag{6.7}$$

(or its rearranged equivalent problem using $\overline{A}^c$ to take advantage of banding of $\overline{A}^c$) for W^{n+1}_1 and W^{n+1}_b. We have simply inverted (6.3) in an efficient and vectorizable manner. Then, using W^{n+1}_b, we complete the block Gauss-Seidel analogy on (6.2) and obtain W^{n+1}_r by solving

$$A_{rr}W^{n+1}_r = Q_4 - A_{rb}W^{n+1}_b. \tag{6.8}$$

Finally, from (6.5), we set

$$\tilde{P}^{n+1} = \begin{pmatrix} P^n_1 \\ P^n_b \\ 0 \\ P^n_r \end{pmatrix} + \tau \begin{pmatrix} W^n_1 \\ W^n_b \\ 0 \\ W^n_r \end{pmatrix}.$$

Since this algorithm only requires two separate solutions of mixed problems on the subregions (each subregion problem possibly being solved via a different parallel processor) and one solution on the original, uniform coarse grid, it is relatively easy to perform. Similarly, no complex data structure is required, and the algorithm can be implemented in existing large-scale codes without severely disrupting the solution process. Promising numerical results for the algorithm have appeared [5,31]. These results have also been extended to more general reservoir simulation problems in a paper by Ewing, Boyett, Babu, and Heinemann [25].

As stated, the algorithm in its most general form involves two separate solutions on the subregions at each step. This iterative procedure is uniformly well-conditioned for finite element procedures such as those used in this paper or point-centered finite difference methods, but not for cell-centered finite differences [31]. For discretizations arising from cell-centered finite difference methods, a scaling of the iteration via the parameter τ in (6.5) may be necessary. The use of the algorithm as a preconditioner [5] for another iterative procedure such as conjugate gradient also involves two distinct solutions on the subregions at each step. This comes from the desire to have a symmetric preconditioner of the form $B(\cdot,\cdot)$ which is important for conjugate gradient methods. As has been pointed out in [49], the FAC algorithm [50] involves only one subregion solution per iteration. See Mandel and McCormick [49] for a comparison of

FAC, the symmetric BEPS [5] preconditioner and the algorithm presented here and their theories.

By considering the domain decomposition techniques presented by Bramble, Pasciak, and Schatz [6] that led to this algorithm, we can see that if the subregion problems ((6.4) and its sequels with updated guesses for P_b^n) are solved exactly, then Q_4^n in (6.6) and (6.7) is identically zero and the iterative method presented here requires only one subregion solution per iteration (from (6.8)). Preliminary computations indicate that if the subregion problem is solved iteratively with its own preconditioner, the full algorithm with two subregion solves will converge faster than the version with one subregion solve for some problems. Iterative solution of the unrefined region causes no difficulty with either version of the algorithm. This is an important consideration for the reservoir simulation applications when iterative solution of the unrefined problem is essential due to their size, since direct solution of the refined region problems is usually not possible.

Adaptive grid-refinement techniques utilizing patch techniques have been presented in several surveys (e.g., see [22-24]). The patch techniques have been incorporated efficiently in existing multiphase industrial reservoir simulation codes. Results for the SPE Comparison projects number 1 and 2 were presented in [25]. The local refinement was both efficient and effective since excellent results were obtained without destroying the efficiency of the original codes.

Acknowledgements

This research was supported in part by National Science Foundation Grant No. DMS-8504360, by Office of Naval Research Contract No. 0014-88-K-0370, by the Institute for Scientific Computation through NSF Grant No. RII-8610680, and by the Pittsburgh and Minnesota Supercomputing Centers.

References

[1] M.B. Allen, III, R.E. Ewing, and P. Lu, "Well conditioned iterative schemes for mixed finite element methods of porous-media flows", *SIAM J. Sci. Stat. Comp.*, submitted.

[2] A.A. Baker, L.W. Gelhar, A.L. Gutjahr, and J.R. Macmillan, 1978, "Stochastic analysis of spatial variability in subsurface flows, I, Comparison of one- and three-dimensional flows", *Water Resour. Res.*, 14 (2), pp. 263-271.

[3] J.W. Barrett and K.W. Morton, 1984, "Approximate symmetrization and Petrov-Galerkin methods for diffusion-convection problems", *Comp. Meth. in Appl. Mech. and Eng.*, 45, pp. 97-122.

[4] A.L. Benham and R.W. Olson, June 1963, "A model study of viscous fingering", *Soc. Pet. Eng. J.*, pp. 138-144.

[5] J.H. Bramble, R.E. Ewing, J.E. Pasciak, and A.H. Schatz, 1988, "A preconditioning technique for the effcient solution of problems with local grid refinement", *Computer Methods in Applied Mechanics and Engineering*, 67, pp. 149-159.

[6] J.H. Bramble, J.E. Pasciak, and A.H. Schatz, 1986, "An iterative method for elliptic problems on regions partitioned into substructures", *Math Comp.*, 46, pp. 361-370.

[7] E.S. Carlson, 1986, "Velocity distributions and overall mobility: their roles in the initiation and propagation of viscous fingering in first contact miscible systems", *Ph.D. Thesis*, University of Wyoming.

[8] M.A. Celia, I. Herrera, E. Bouloutas, and J.S. Kindred, 1989, "A new numerical approach for the advective-diffusive transport equation", *Numerical Methods for PDE's*, 5, pp. 203-226.

[9] M.A. Celia, T.F. Russell, I. Herrera, and R.E. Ewing, in press, "An Eulerian-Lagrangian localized adjoint method for the advection-diffusion equation", *Advances in Water Resources.*

[10] C. Chardair-Riviere, G. Chavert, and J. Jaffré, 1990, "Multiscale representation for simultaneous estimation of relative permeabilities and capillary pressure, *SPE*, 20501, Proc. 65th Annual Technical Conf. SPE, New Orleans, Louisiana, September 23–26.

[11] G. Chavent, 1976, "A new formulation of diphasic incompressible flows in porous media", *Lecture Notes in Mathematics*, No. 503, Springer-Verlag.

[12] R.L. Chuoke, P. Van Meurs, and C. Van Der Poel, 1959, "The instability of slow, immiscible, viscous, liquid-liquid displacements in permeable media", *Trans. AIME*, 216, pp. 188-194.

[13] H.K. Dahle, M.S. Espedal, and R.E. Ewing, 1988, "Characteristic Petrov-Galerkin subdomain methods for convection diffusion problems", *IMA Volume 11, Numerical Simulation in Oil Recovery*, M.F. Wheeler, ed., Springer-Verlag, Berlin, pp. 77-88.

[14] H.K. Dahle, M.S. Espedal, R.E. Ewing, and O. Saevareid, in press, "Characteristic adaptive sub-domain methods for reservoir flow problems", *Numerical Methods for PDE's.*

[15] L. Demkowitz and J.T. Oden, 1986, "An adaptive characteristic Petrov-Galerkin finite element method for convection-dominated linear and nonlinear parabolic problems in two space variables", *Comp. Meth. in Appl. Mech. and Eng.*, 55, pp. 63-87.

[16] J. Douglas, Jr., R.E. Ewing, and M.F. Wheeler, 1983, "The approximation of the pressure by a mixed method in the simulation of miscible displacement", *R.A.I.R.O. Analyse Numerique*, 17, pp. 17-33.

[17] J. Douglas, Jr., R.E. Ewing, and M.F. Wheeler, 1983, "A time-discretization procedure for a mixed finite element approximation of miscible displacement in porous media", *R.A.I.R.O. Analyse Numerique*, 17, pp. 249-265.

[18] J. Douglas, Jr., and T.F. Russell, 1982, "Numerical methods for convection dominated diffusion problems based on combining the modified method of characteristics with finite element or finite difference procedures", *SIAM J. Numer. Anal.*, 19, pp. 871–885.

[19] M.S. Espedal and R.E. Ewing, 1987, "Characteristic Petrov-Galerkin subdomain methods for two-phase immiscible flow", *Comp. Meth. Appl. Mech. and Eng.*, 64, pp. 113-135.

[20] R.E. Ewing, 1982, "Determination of coefficients in reservoir simulation", *Numerical Treatment of Inverse Problems for Differential and Integral Equations*, P. Deuflhardt and E. Hairer, eds., Birkhauser, Berlin, pp. 206-226.

[21] R.E. Ewing, 1983, "Problems arising in the modeling of processes for hydrocarbon recovery", in *The Mathematics of Reservoir Simulation*, R.E. Ewing, ed., Frontiers in Applied Mathematics, Vol. 1, SIAM, Philadelphia, pp. 3-34.

[22] R.E. Ewing, 1986, "Adaptive mesh refinement in large-scale fluid flow simulation", *Accuracy Estimates and Adaptivity for Finite Elements*, Ch. 16 (I. Babuska, O. Zienkiewicz, and E. Oliveira, eds.), John Wiley and Sons, New York, pp. 299-314.

[23] R.E. Ewing, 1986, "Efficient adaptive procedures for fluid flow applications", *Comp. Meth. Appl. Mech. Eng.*, 55, pp. 89-103.

[24] R.E. Ewing, 1987, "Adaptive grid refinement methods for time-dependent flow problems", *Comm. Appl. Num. Meth.*, 3, pp. 351-358.

[25] R.E. Ewing, B.A. Boyett, D.K. Babu, and R.F. Heinemann, 1989, "Efficient use of locally refined grids for multiphase reservoir simulation", *SPE 18413, Proceedings, Tenth SPE Symposium on Reservoir Simulation*, Houston, Texas, February 6-8, pp. 55-70.

[26] R.E. Ewing and J.H. George, 1984, "Viscous fingering in hydrocarbon recovery processes", *Mathematical Methods in Energy Research*, K.I. Gross, ed., SIAM, Philadelphia, pp. 194-213.

[27] R.E. Ewing and J.H. George, 1985, "Identification and control of distributed parameters in porous media flow", *Distributed Parameter Systems*, Lecture Notes in Control and Information Sciences (M. Thoma, ed.), Springer-Verlag, 70, May, pp. 145-161.

[28] R.E. Ewing and R.F. Heinemann, 1983, "Incorporation of mixed finite element methods in compositional simulation for reduction of numerical dispersion", *Proc. Seventh SPE Symposium on Reservoir Simulation*, SPE No. 12267, San Francisco, November 15-18, pp. 341-347.

[29] R.E. Ewing and R.F. Heinemann, 1984, "Mixed finite element approximation of phase velocities in compositional reservoir simulation", *Computer Meth. Appl. Mech. Eng.*, R.E. Ewing, ed., 47, pp. 161-176.

[30] R.E. Ewing, J.V. Koebbe, R. Gonzalez, and M.F. Wheeler, 1985, "Mixed finite element methods for accurate fluid velocities", *Finite Elements in Fluids*, Vol. 4, John Wiley, pp. 233-249.

[31] R.E. Ewing, R.D. Lazarov, and P.S. Vassilevski, in press, "Local refinement techniques for elliptic problems on cell-centered grids, I. Error analysis and II. Optimal-order two-grid iterative methods", *Math Comp.*

[32] R.E. Ewing, T.F. Russell, and M.F. Wheeler, 1983, "Simulation of miscible displacement using mixed methods and a modified method of characteristics", *Proc. Seventh SPE Symposium on Reservoir Simulation*, SPE No. 12241, San Francisco, November 15-18, pp. 71-82.

[33] R.E. Ewing, T.F. Russell, and M.F. Wheeler, 1984, "Convergence analysis of an approximation of miscible displacement in porous media by mixed finite elements and a modified method of characteristics", *Computer Meth. Appl. Mech. Eng.*, R.E. Ewing, ed., 47, pp. 73-92.

[34] R.E. Ewing, T.F. Russell, and L.C. Young, 1989, "An anisotropic coarse-grid dispersion model of heterogeneity and viscous fingering in five-spot miscible displacement that matches experiments and fine-grid simulations", *SPE 18441, Proceedings, 10th SPE Reservoir Simulation Symposium*, Houston, Texas, pp. 447-466.

[35] R.E. Ewing and M.F. Wheeler, 1983, "Computational aspects of mixed finite element methods", in *Numerical Methods for Scientific Computing*, R.S. Stepleman, ed., North-Holland Publishing Co., pp. 163-172.

[36] T.P. Fishlock and W.R. Rodwell, 1983, "Improvements in the numerical simulation of carbon dioxide displacement", European Paris Conference.

[37] J.W. Gardner and J.G.J. Ypma, "Investigation of phase behavior-macroscopic bypassing interaction in CO_2 flooding", SPE 10686.

[38] L.W. Gelhar and C.L. Axness, 1983, "Three-dimensional stochastic analysis of macro-dispersion in aquifers", *Water Resour. Res.*, 19 (1), pp. 161-180.

[39] J.M. Hagoort, 1974, "Displacement stability of water drives in water-wet connate-water-bearing reservoirs", *Soc. Pet. Eng. J.*, February, pp. 63-74.

[40] J.P. Heller, 1966, "Onset of instability patterns between miscible fluids in porous media", *J. Appl. Physics*, 37, pp. 1566-1579.

[41] I. Herrera, 1985, "Unified formulation of numerical methods I. Green's formula for operators in discontinuous fields", *Numerical Methods in PDE's*, 1, pp. 25-44.

[42] I. Herrera and R.E. Ewing, 1989, "Localized adjoint methods: application to multiphase flow problems", *Proceedings of the Fifth Wyoming Enhanced Oil Recovery Symposium*, Casper, Wyoming, May 10-11, pp. 147–166.

[43] I. Herrera, R.E. Ewing, M.A. Celia, and T.F. Russell, "Eulerian-Lagrangian localized adjoint methods", (in preparation).

[44] G.R. Jerauld, H.T. Davis, and L.E. Scriven, 1984, "Frontal structure and stability in immiscible displacement", SPE/DOE Fourth Symposium on Enhanced Oil Recovery 2, pp. 135-144.

[45] E.J. Koval, 1963, "A method for predicting the performance of unstable miscible displacement in heterogeneous media", *Soc. Pet. Eng. J.*, June, pp. 145-154.

[46] L.W. Lake and G.J. Hirasaki, 1981, "Taylor's dispersion in stratified porous media", *Soc. Pet. Eng. J.*, 21, pp. 459-468.

[47] R.G. Larson, H.T. Davis, and L.E. Scriven, 1981, "Displacement of residual nonwetting fluid from porous media", *Chem. Eng. Sci.*, 36, pp. 75-85.

[48] T. Lin and R.E. Ewing, 1986, "Parameter estimation for distributed systems arising in fluid flow problems via time series methods", *Proceedings of Conference on "Inverse Problems"*, Oberwolfach, West Germany, Birkhauser, Berlin, 77, pp. 117-126.

[49] J. Mandel and S. McCormick, 1989, "Iterative solution of elliptic equations with refinement: the two-level case," Domain Decomposition Methods (T.F. Chan, R. Glowinski, J. Periaux, and O. Widlund, eds.), SIAM Publications, Philadelphia, Pennsylvania, pp. 81–92.

[50] S. McCormick and J. Thomas, 1986, "The fast adaptive composite grid method for elliptic boundary value problems," *Math Comp.*, 46, pp. 439–456.

[51] A.H. Nayfeh, 1972, "Stability of liquid interfaces in porous media", *Physics of Fluids*, 15 (10), pp. 1751-1754.

[52] S.G. Oh and J.C. Slattery, 1979, "Interfacial tension required for significant displacement of residual oil", *Soc. Pet. Eng. J.*, 19, pp. 83.

[53] H.D. Outmans, 1962, "Transient interfaces during immiscible liquid-liquid displacement in porous media", *Soc. Pet. Eng. J.*, June, pp. 156-164.

[54] H.D. Outmans, 1962, "Nonlinear theory for frontal stability and viscous fingering in porous media", *Soc. Pet. Eng. J.*, June, pp. 165-176.

[55] H.D. Outmans, 1963, "On unique solutions for steady-state fingering in a porous medium", *J. Geophysical Research*, 68, pp. 5735-5737.

[56] C.W. Park and G.M. Homsy, in press, "Two-phase displacement in Hele-Shaw cells: theory", *J. Fluid Mech.*

[57] T.K. Perkins and O.C. Johnston, 1963, "A review of diffusion and dispersion in porous media", *Soc. Pet. Eng. J.*, 3, pp. 70-84.

[58] T.K. Perkins, O.C. Johnston, and R.N. Hoffman, 1965, "Mechanics of viscous fingering in miscible systems", *Soc. Pet. Eng. J.*, December, pp. 301-317.

[59] T.K. Perkins and O.C. Johnston, 1969, "A study of immiscible fingering in linear models", *Soc. Pet. Eng. J.*, March, pp. 39-46.

[60] R.L. Perrine and A.M. Gay, 1966, "Unstable miscible flow in heterogeneous systems", *Soc. Pet. Eng. J.*, September, pp. 228-238.

[61] R.L. Perrine, 1961, "Stability theory and its uses to optimize solvent recovery of oil", *Soc. Pet. Eng. J.*, March, pp. 9-16.

[62] E.J. Peters and D.L. Flock, 1981, "The onset of instability during two-phase immiscible displacement in porous media", *Soc. Pet. Eng. J.*, April, pp. 249-258.

[63] E. Pitts, 1980, "Penetration of fluid into a Hele-Shaw cell: the Saffman-Taylor experiment", *J. Fluid Mech.*, 97, pp. 53-64.

[64] H.H. Rachford, Jr., 1964, "Instability in water flooding oil from water-wet porous media containing connate water", *Soc. Pet. Eng. J.*, June, pp. 133-148.

[65] T.F. Russell, 1985, "The time-stepping along characteristics with incomplete iteration for Galerkin approximation of miscible displacement in porous media", *SIAM J. Numer. Anal.*, 22, pp. 970-1013.

[66] T.F. Russell, 1989, "Eulerian-Lagrangian localized adjoint methods for advection-dominated problems", *Proceedings of 13th Biennial Conference on Numerical Analysis*, Pitmann Publishing Company, Dundee, Scotland, June 27-30.

[67] T.F. Russell and M.F. Wheeler, 1983, "Finite element and finite difference methods for continuous flows in porous media", in *The Mathematics of Reservoir Simulation*, (R.E. Ewing, ed.) *Frontiers in Applied Mathematics*, SIAM, Philadelphia.

[68] P.G. Saffman, 1982, "Fingering in porous media", Lecture Notes in Physics *154*, R. Burridge, ed., pp. 208-215.

[69] P.G. Saffman and G. Taylor, 1958, "The penetration of a fluid into a porous medium or Hele-Shaw cell containing a more viscous liquid", *Proc. Roy. Soc. A*, 245, pp. 312-329.

[70] P.G. Saffman, 1959, "Exact solutions for the growth of fingers from a flat interface between two fluids in a porous medium or Hele-Shaw cell", *Quart. J. Mech. and Appl. Math.*, 12, pp. 146-150.

[71] A.E. Scheidegger, 1960, "General spectral theory for the onset instabilities in displacement processes in porous media", *Geof. Pura Appl.*, 47, pp. 41.

[72] A.E. Scheidegger, 1960, "Growth of instabilities on displacement fronts in porous media", *Physics of Fluids*, 3, pp. 94.

[73] J.C. Slattery, 1969, "Single-phase flow through porous media", *AIChE J.*, 15, pp. 866-872.

[74] J.C. Slattery, 1970, "Two-phase flow through porous media", *AIChE J.*, 16, pp. 345-352.

[75] G.I. Taylor, 1953, "Dispersion of soluble matter in solvent flowing slowly through a tube", *Proc. Roy. Soc.*, A219, pp. 183-203.

[76] G.I. Taylor, 1954, "Conditions under which dispersion of a solute in a stream of solvent can be used to measure molecular diffusion", *Proc. Roy. Soc.*, A225, pp. 473-477.

[77] M.R. Todd and W.J. Longstaff, 1972, "The development, testing and application of a numerical simulator for predicting miscible flood performance", *Jour. Pet. Tech.*, 253, pp. 874-882.

[78] P. Van Meurs and C. Van Der Poel, 1958, "A theoretical description of water-drive processes involving viscous fingering", *Trans. AIME*, 213, pp. 103-112.

[79] J.M. Vanden-Broeck, 1983, "Fingers in a Hele-Shaw cell with surface tension", *Physics of Fluids*, 8, p. 26.

[80] J.E. Warren and F.F. Skiba, 1964, "Macroscopic dispersion", *Soc. Pet. Eng. J.*, 4, pp. 215-230.

[81] R.A. Wooding, 1962, "The stability of an interface between miscible fluids in a porous medium", *Zeit. Fur Ang. Math. Und Physik*, 13, pp. 255-265.

[82] Y.C. Yortsos and A.B. Huang, 1984, "Linear stability of immiscible displacement including continuously changing mobility and capillary effects", SPE/DOE Fourth Symposium on Enhanced Oil Recovery 2, pp. 145-162.

[83] L.C. Young, 1984, "A study of spatial approximations for simulating fluid displacements in petroleum reservoirs", *Comp. Meth. in Appl. Mech. Eng.*, 47, pp. 3-46.

Transport in Porous Media **6**: 501–517, 1991
© 1991 *Kluwer Academic Publishers. Printed in the Netherlands.*

Geometric properties of two phase flow in geothermal reservoirs

GRAHAM J WEIR

Applied Mathematics Group, Physical Sciences Division, Department of Scientific and Industrial Research, PO Box 1335, Wellington, New Zealand.

(Received)

Abstract The two phase flow equations frequently used in geothermal engineering ignore capillary pressure, which results in a singular system of equations. Analysis of these equations reveals three mechanisms for altering saturation : local boiling, the spatial dependence of flowing enthalpy due to the convective transport of fluid, and counterflow. A scalar function is associated with each of these three mechanims. At each point in space, flows are essentially two dimensional, with gravity establishing a vertical hierarchy, in that volumetric, energy and mass fluxes can never point below a lower member in this triple. With increasing liquid saturation, the characteristics associated with the saturation equation move up from below this grouping of directions, and eventually may even point above volumetric fluxes. Finally, weak shocks and the associated entropy condition are considered. The characteristics of the saturation equation coincide with the velocity of extremely weak shocks, and saturation increases with the passage of a weak shock, provided the magnitude of the characteristic speed increases with saturation.

Key words. Two phase, local boiling, flowing enthalpy, counterflow, singular equations, pressure, saturation, vertical hierarchy, characteristics, weak shocks, theory.

0. Notation

Table 1

Symbol	Meaning	Estimate (MKS)
C_l	liquid heat capacity	4.6×10^3
C_m	rock heat capacity	9×10^2
C_v	vapour heat capacity	-5×10^2
G	counterflow energy flux	0 -1
h	flowing enthalpy	$(1 - 2.8) \times 10^6$
h_l	liquid enthalpy	$1. \times 10^6$
h_v	vapour enthalpy	2.8×10^6
k	permeability	1×10^{-16}
$\mathbf{k}$	downward vector	
P	pressure	3.9776×10^6
S	liquid saturation	$0 - 1.$
T	temperature	250
dT/dP	derivative at saturation	1.5×10^{-5}
z	vertical coordinate	
μ_l	liquid viscosity	1.1×10^{-4}
μ_v	vapour viscosity	1.7×10^{-5}
ρ_l	liquid density	8×10^2
ρ_m	rock density	2.65×10^3
ρ_v	vapour density	2×10^1
ϕ	porosity	0.2
$\mathcal{E}_{\mathcal{P}}$	-	3×10^1
$\mathcal{E}_{\mathcal{S}}$	-	$(7 - 300) \times 10^6$

1. Introduction

The main aim of this paper is to present some new and general results, applicable to the the equations of two phase flow, as formulated in geothermal reservoir engineering. Two phase regions are important in many geothermal reservoirs, especially at depths of order several hundred metres, where rising, essentially isothermal single phase liquid first begins to boil. The fluid then continues to rise, with its temperature and pressure closely following the saturation (boiling) curve appropriate to the fluid composition.

Perhaps the two most interesting theoretical aspects of the (idealised) two phase flow equations in geothermal reservoir engineering are that firstly, only one component (water) is involved; and secondly, that the densities of the two phases are so different. This has led to the approximation of ignoring capillary pressure. The main aim of this paper is to analyse some of the consequences of this assumption, especially in relation to saturation changes within a uniform porous medium.

A general analytic treatment of three dimensional flow is considered. Previously, three dimensional modelling in geothermal reservoirs have relied on numerical simulators. In contrast, most of the past analytic work has been restricted to one dimensional examples.

Perhaps the most general, published derivation of the two phase equations in geothermal engineering is that by Garg and Pritchett (1977), who include an additional term resulting from the interaction between temperature and the potential energy of the fluid. This term usually has a small effect in geothermal reservoirs, and its omission leads to the equations used in this paper.

We also ignore thermal conduction in this paper, largely because the inclusion of conduction usually produces only small changes to the physical variables of pressure and saturation. In addition, we anticipate no new phenomena appearing from the inclusion of conductivity in the energy equation, since in a two phase region, where temperature and pressure are functionally related, a temperature gradient implies a pressure gradient. Consequently, conductivity can be considered simply to modify the apparent enthalpy of the fluid in the energy equation.

This paper begins by outlining the equations used in this paper. The singular nature of these equations is emphasised, and the characteristic velocity of the saturation equation is derived. The relationship between these characteristics and weak discontinuities (shocks) in saturation is established. Some of the geometric properties of weak shocks are derived. Volumetric fluxes are shown to be continuous across a shock, which allows a calculus of saturation derivatives to be formulated. Finally, the relative orientation in the vertical of mass, energy and volumetric fluxes is established, together with the corresponding orientation of the characteristics.

2. Equations

The equations most commonly used in geothermal reservoir engineering (Grant *et al.*, 1982) to describe conservation of mass and energy in a uniform porous are respectively:

$$\frac{\partial M}{\partial t} = \nabla \cdot (F_l(\nabla P - \rho_l g\mathbf{k}) + F_v(\nabla P - \rho_v g\mathbf{k})) \tag{1}$$

$$= \nabla \cdot (F_M \nabla P - G_M \mathbf{k}) \tag{2}$$

$$\frac{\partial E}{\partial t} = \nabla \cdot (h_l F_l(\nabla P - \rho_l g\mathbf{k}) + h_v F_v(\nabla P - \rho_v g\mathbf{k})) \tag{3}$$

$$= \nabla \cdot (F_E \nabla P - G_E \mathbf{k}) \tag{4}$$

where

$$M = \phi S \rho_l + \phi(1-S)\rho_v, \tag{5}$$

$$E = (1-\phi)\rho_m C_m T + \phi S \rho_l U_l + \phi(1-S)\rho_v U_v,$$

$$= (1-\phi)\rho_m C_m T + \phi S \rho_l h_l + \phi(1-S)\rho_v h_v - \phi P, \tag{6}$$

$$F_l = k k_{rl} \rho_l / \mu_l, \tag{7}$$

$$F_v = k k_{rv} \rho_v / \mu_v \tag{8}$$

Mass and energy fluxes are respectively

$$\mathbf{J}_M = -F_M \nabla P + G_M \mathbf{k} \tag{9}$$

$$\mathbf{J}_E = -F_E \nabla P + G_E \mathbf{k} \tag{10}$$

where the corresponding flux coefficients are

$$F_M = F_l + F_v, \tag{11}$$
$$F_E = h_l F_l + h_v F_v, \tag{12}$$
$$G_M = (\rho_l F_l + \rho_v F_v) g, \tag{13}$$
$$G_E = (h_l \rho_l F_l + h_v \rho_v F_v) g, \tag{14}$$
$$h = F_E / F_M \tag{15}$$

In (1 - 15), M and E are respectively mass and energy densities, P pressure, t time, $\mathbf{k}$ unit vector, directed vertically downwards, ∇ is the gradient operator, $\nabla\cdot$ is the divergence operator, ϕ porosity, S liquid saturation, ρ_l and ρ_v are respectively liquid and vapour density, ρ_m dry rock density, C_m rock specific heat, T temperature, U_l and U_v internal energy for liquid and vapour, k permeability, k_{rl} and k_{rv} the permeability reduction factors for respectively liquid and vapour, μ_l and μ_v the dynamic viscosity of respectively liquid and vapour, g is gravitational acceleration and h is flowing enthalpy, which is independent of porosity and permeability.

Non-zero ranges of saturation exist for which one phase is immobile. We assume there are residual saturations, S_* and S^*, such that

$$k_{rv} = 0 \qquad S^* < S < 1 \tag{16}$$
$$k_{rl} = 0 \qquad 0 < S < S_* \tag{17}$$

In a region of two phase fluid, we define a 'liquid flow' region to exist for saturations above S^*, a 'vapour flow' region to exist for saturations below S_*, and a mixed flow region to exist for saturations between S_* and S^*. Single phase regions can also be important in discussions of two phase regions, since sometimes two phase conditions cannot be preserved everywhere, especially when shocks occur. However, in this paper, we shall not discuss such single phase regions.

Consequently, for the remainder of this paper, we assume that fluid satisfies the boiling condition,

$$P = P_{\text{sat}}(T)$$

or namely, that pressure equals the saturation pressure for the corresponding temperature.

The two second order equations in (1) and (3) can be written as a single first order 'pressure equation' (Burnell *et al.*, 1989) by eliminating the time derivatives of saturation, and as a first order 'saturation equation', by eliminating second order spatial derviatives. This paper shall not discuss the 'pressure equation', but shall concentrate on the 'saturation equation',

$$\frac{\partial E}{\partial t} - h\frac{\partial M}{\partial t} = -\mathbf{J}_M \cdot \nabla h + \frac{\partial G}{\partial z}$$

which can be rewritten as

$$-\mathcal{E}_S\frac{\partial S}{\partial t} = \mathcal{E}_P\frac{\partial P}{\partial t} + \mathbf{J}_M \cdot \nabla h - \frac{\partial G}{\partial z} \tag{18}$$

where z is the coordinate for the $\mathbf{k}$ direction, and

$$\mathcal{E}_P = \frac{\partial E}{\partial P} - h\frac{\partial M}{\partial P}, \tag{19}$$

$$\mathcal{E}_S = \frac{\partial E}{\partial S} - h\frac{\partial M}{\partial S}. \tag{20}$$

$$G = hG_M - G_E \tag{21}$$

We shall call (18) the 'saturation equation', since if the pressure field is known, and suitable boundary and initial conditions are known for saturation, then (18) can be integrated to yield saturation.

Estimates for $\mathcal{E}_P$, $\mathcal{E}_S$ and G follow from differentiating (5) and (6) with respect to saturation and pressure, and using (11) to (15),

$$\mathcal{E}_P \approx [(1-\phi)\rho_m C_m + \phi S \rho_l \mathcal{C}_l + \phi(1-S)\rho_v \mathcal{C}_v]\frac{dT}{dP} \tag{22}$$

$$-\mathcal{E}_S = \phi(h_v - h_l)[F_v\rho_l + F_l\rho_v]/F_M \tag{23}$$

$$G = gF_lF_v(h_v - h_l)(\rho_l - \rho_v)/F_M \tag{24}$$

where

$$\mathcal{C}_l = \frac{dh_l}{dT} \tag{25}$$

$$\mathcal{C}_v = \frac{dh_v}{dT} \tag{26}$$

In (25) and (26), $\mathcal{C}_l$ and $\mathcal{C}_v$ are 'apparent' heat capacities along the saturation line. The total derivative with respect to temperature arises, since along the saturation line, both temperature and pressure vary, and so these apparent heat capacities do not correspond to the standard heat capacites for liquid and vapour.

The saturation equation, (18), shows there are three mechanisms for altering saturation. The first term on the right hand side of (18) relates saturation changes to local changes in pressure (temperature), and so corresponds to local boiling. This mechanism is important for example in linearised horizontal flows. The second term on the right hand side of (18) follows from the non-constancy of flowing enthalpy, and operates since the pressure gradient driving a mass flux implies that flowing enthalpy varies with position, and so there is implicitly an energy imbalance associated with a mass flux. This expression is nonlinear in pressure, and has been analysed (Burnell *et al.*, 1989) in relation to boundary layer formation. The last term on the right hand side of (18) dominates when temporal changes in pressure are unimportant, and there is no mass flux. This corresponds to steady counterflow (Mc Guinness and Pruess, 1987) conditions, when equal masses of fluid are descending and ascending.

Each expression in (18) has dimensions of energy flux per distance, or equivalently energy density per time. Consequently, the scalar function G corresponds to an energy flux expression, and is associated with counterflow. Indeed, G equals the energy flux in steady vertical counterflows, when the net mass flux is zero. Similarly, the scalar function of flowing enthalpy h can be associated with the second method for altering saturation. Note that h does not always correspond to the exact proportionality between energy and mass fluxes, since in general these two vectors are non- collinear. Finally, consideration of local boiling suggests the scalar function E - $\int hdM$ may be important, which has been established (Burnell *et al.*, 1990) for some horizontal flows.

When both phases are mobile (a mixed flow region), it may be convenient to expand the gradients of h and G in (18) in terms of gradients of P and S to yield

$$-\mathcal{E}_S(\frac{\partial S}{\partial t}+\mathbf{c}\cdot\nabla S)=\mathcal{E}_P\frac{\partial P}{\partial t}-F_M\frac{\partial h}{\partial P}\nabla P\cdot\nabla P+\frac{\partial P}{\partial z}(\frac{\partial G_E}{\partial P}-h\frac{\partial G_M}{\partial P}), \quad (27)$$

showing that saturation satisfies a first order partial differential equation, with characteristic velocity, $\mathbf{c}$, satisfying

$$\mathcal{E}_S\mathbf{c}=\frac{\partial h}{\partial S}\mathbf{J}_M-\frac{\partial G}{\partial S}\mathbf{k} \quad (28)$$

3. Weak Shocks

The main aim of this section is to investigate the physical relevance of the characteristic velocity $\mathbf{c}$ introduced in (28). Since first order nonlinear equations often contain shock fronts, we shall begin by investigating weak shocks, in which any discontinuity in saturation is sufficiently small that only terms linear in the saturation discontinuity need be retained. Across any shock front, pressure is assumed to be continuous. The discontinuities at a shock result, then, from discontinuities of saturation and discontinuities in gradients of pressure.

Suppose a shock occurs, and moves with velocity **b**. Conservation of mass and energy at the shock then requires, from (5) and (6),

$$\mathbf{J}_M^u - \mathbf{J}_M^d = \phi(\rho_l - \rho_v)(S^u - S^d)\mathbf{b} \tag{29}$$

$$\mathbf{J}_E^u - \mathbf{J}_E^d = \phi(h_l\rho_l - h_v\rho_v)(S^u - S^d)\mathbf{b} \tag{30}$$

where superscripts u and d denote upstream and downstream values across a shock.

Eliminating **b** from (29) and (30) shows that the vector of volumetric flux, $\mathbf{J}_Q$, is continuous across a shock, since from (1), (9) and (10),

$$\begin{aligned}\rho_l\rho_v(h_v - h_l)\mathbf{J}_Q &= \rho_l\rho_v(h_v - h_l)\left[- F_l(\nabla P - \rho_l g\mathbf{k})/\rho_l - F_v(\nabla P - \rho_v g\mathbf{k})/\rho_v\right] \\ &= (\rho_l - \rho_v)\mathbf{J}_E - (h_l\rho_l - h_v\rho_v)\mathbf{J}_M \end{aligned}\tag{31}$$

Continuity of volumetric fluxes across a shock allows a partial differentiation of vector fluxes with respect to saturation to be formulated, by requiring that first order changes in vector fluxes and saturation are chosen so as to preserve the continuity of volumetric fluxes.

Firstly, we shall choose as a basis, mass flux $\mathbf{J}_M$ and the vertical vector **k**, and so from (9), (10), (15) and (21), energy fluxes can be reexpressed as

$$\mathbf{J}_E = h\mathbf{J}_M - G\mathbf{k} \tag{32}$$

Then, from the continuity of $\mathbf{J}_Q$,

$$\mathcal{E}_S^d(\mathbf{J}_M^u - \mathbf{J}_M^d) = (\mathcal{E}_S^d - \mathcal{E}_S^u)\mathbf{J}_M^u - \phi(\rho_l - \rho_v)(G^u - G^d)\mathbf{k} \tag{33}$$

which suggests adopting the notation

$$\frac{\partial \mathbf{J}_M}{\partial S} = \frac{-1}{\mathcal{E}_S}\frac{\partial \mathcal{E}_S}{\partial S}\mathbf{J}_M - \frac{\phi(\rho_l - \rho_v)}{\mathcal{E}_S}\frac{\partial G}{\partial S}\mathbf{k} \tag{34}$$

since to first order terms in $(S^u - S^d)$,

$$\mathbf{J}_M^u = \mathbf{J}_M^d + (S^u - S^d)\frac{\partial \mathbf{J}_M}{\partial S} \tag{35}$$

where partial derivatives are evaluated at the downstream values.

From (23) and (15), partial derivatives of h and $\mathcal{E}_S$ with respect to saturation are interchangable, since

$$\mathcal{E}_S + \phi(\rho_l - \rho_v)h = \phi(\rho_l h_l - \rho_v h_v) \tag{36}$$

Equation (36) allows the shock and characteristic velocities to be related to within first order saturation changes, since from (29) and (30),

$$\mathbf{b} = \mathbf{c}^d + \frac{1}{2}(S^u - S^d)\frac{\partial \mathbf{c}}{\partial S} \tag{37}$$

$$\frac{1}{\mathcal{E}_S}\frac{\partial \mathbf{c}}{\partial \mathrm{S}} = \frac{\partial}{\partial S}(\frac{1}{\mathcal{E}_S^2}\frac{\partial h}{\partial S})\mathbf{J}_M - \frac{\partial}{\partial S}(\frac{1}{\mathcal{E}_S^2}\frac{\partial G}{\partial S})\mathbf{k} \tag{38}$$

and so for very weak shocks, the characteristic and shock directions coincide. Additionally, to first order, the shock direction bisects the angle between the two intersecting characteristics. The results above on saturation derivatives are perhaps obtained most easily by expressing all vector quantities as a weighted sum of $\mathbf{J}_M$ and $\mathbf{k}$, and noting that the saturation partial derivatives behave as ordinary differentiations, provided (34) is used.

Finally, an 'entropy condition' holds at a shock, since there characteristics upstream of the shock are running into the shock, and those downstream are being overtaken by the shock, yielding the relationship

$$\mathbf{b} \cdot \mathbf{c}^d \; < \; \mathbf{b} \cdot \mathbf{b} \; < \mathbf{b} \cdot \mathbf{c}^u \tag{39}$$

which, for weak shocks, reduces to

$$(S^u - S^d)\frac{\partial \mathbf{c} \cdot \mathbf{c}}{\partial S} \; > \; 0 \tag{40}$$

Consequently, provided the magnitude of the shock velocity increases with saturation at a shock, then so does the jump in saturation, $S^u - S^d$. For weak shocks, characteristic magnitudes satisfy, to first order,

$$\mathbf{c}^d \cdot \mathbf{c}^u = \mathbf{b} \cdot \mathbf{b} \tag{41}$$

4. Vertical hierarchies

The previous sections have shown, from (32), (31) and (28), that the two phase equations of geothermal reservoir engineering imply that the four vectors $\mathbf{J}_M$, $\mathbf{J}_E$, $\mathbf{J}_Q$ and $\mathbf{c}$ are spanned by the two vectors $\mathbf{J}_M$ and $\mathbf{k}$. Consequently, for each spatial position, $\mathbf{J}_M$, $\mathbf{J}_E$, $\mathbf{J}_Q$ and $\mathbf{c}$ all lie in a vertical plane, and so these three dimensional vectors are essentially only two dimensional.

This section investigates the vertical orientation of the four vectors $\mathbf{J}_M$, $\mathbf{J}_E$, $\mathbf{J}_Q$ and $\mathbf{c}$ within such planes. The mathematical motivation for this follows from (32), which shows that energy fluxes never point below mass fluxes. This follows since the scalar function G is never negative, and is only nonzero for mixed flows. (Indeed, if one phase is immobile, then these vectors are collinear). From a physical viewpoint, there is a natural tendency for vapour to rise relative to liquid, which may impose a vertical ordering on the system.

From (31), (32) and (36), volumetric flux (for water-steam phases) never points below the corresponding energy flux, since

$$\rho_l \rho_v (h_v - h_l) h \mathbf{J}_Q = \frac{-\mathcal{E}_S}{\phi} \mathbf{J}_E - (\rho_l h_l - \rho_v h_v) G \mathbf{k} \tag{42}$$

and so the mass, energy and volumetric fluxes are ordered vertically, with volumetric fluxes directed above the others; energy flux is directed above mass flux. As a corollary, should one flux vector be zero at some point, then the other flux vectors are vertical there.

It remains to check the vertical orientation of the characteristics, $\mathbf{c}$, relative to $\mathbf{J}_M$, $\mathbf{J}_E$ and $\mathbf{J}_Q$. From (28), (31) and (32),

$$-\mathcal{E}_S \mathbf{c} = -\frac{\partial h}{\partial S} \mathbf{J}_M + \frac{\partial G}{\partial S} \mathbf{k} \tag{43}$$

$$-\mathcal{E}_S h \mathbf{c} = -\frac{\partial h}{\partial S} \mathbf{J}_E + h^2 \frac{\partial}{\partial S} \left(\frac{G}{h}\right) \mathbf{k} \tag{44}$$

$$\mathbf{c} = -\frac{\partial h}{\partial S} \frac{\phi \rho_l \rho_v (h_v - h_l)}{\mathcal{E}_S^2} \mathbf{J}_Q + \frac{\partial}{\partial S} \left(\frac{G}{-\mathcal{E}_S}\right) \mathbf{k} \tag{45}$$

and so the vertical orientation of $\mathbf{c}$ is below or above $\mathbf{J}_M$, $\mathbf{J}_E$ and $\mathbf{J}_Q$ whenever the derivative with respect to saturation of respectively G, G/h and $G/-\mathcal{E}_S$ is positive or negative. Recall that $\mathbf{c}$ is nonzero only for mixed flows, so we shall assume for the remainder of this section that both phases are mobile.

We shall show that the three positive functions G, G/h and $G/-\mathcal{E}_S$ all have maxima in the case when

$$k_{rl} + k_{rv} = 1 \tag{46}$$

and when k_{rl} increases with saturation. Equation (46) is frequently assumed in geothermal reservoir calculations.

Calculation shows that, subject to (46), the maxima in G, G/h and $G/-\mathcal{E}_S$ satisfy

$$\frac{\partial G}{\partial S} = 0 \quad \Leftrightarrow \quad (\frac{k_{rv}}{k_{rl}})^2 = \frac{\rho_l \mu_v}{\rho_v \mu_l} \tag{47}$$

$$\frac{\partial}{\partial S}(\frac{G}{h}) = 0 \quad \Leftrightarrow \quad (\frac{k_{rv}}{k_{rl}})^2 = \frac{h_l \rho_l \mu_v}{h_v \rho_v \mu_l} \tag{48}$$

$$\frac{\partial}{\partial S}(\frac{G}{-\mathcal{E}_S}) = 0 \quad \Leftrightarrow \quad (\frac{k_{rv}}{k_{rl}})^2 = \frac{\mu_v}{\mu_l} \tag{49}$$

Three dimensional plots of the three functions G, G/h, and $G/-\mathcal{E}_S$ are given in Figures 1, 2 and 3, as functions of k_{rl} and pressure, subject to (46). G has been scaled relative to the permeability of the medium k, and $\mathcal{E}_S$ relative to medium porosity ϕ, in these figures in order to obtain plots largely independent of the porous medium. The maxima in G, G/h, and $G/-\mathcal{E}_S$ implied by the numerical estimates in Table 1, occur for k_{rl} equalling respectively .3, .4 and .7, in good agreement with Figures 1 to 3.

The vertical orientation of the characteristics can then be inferred from (43) to (49), as tending to increase in a positive vertical direction with increasing saturation. For sufficiently small saturations, the characteristics point below the mass flux vector, whereas for saturations sufficiently large, the characteristics are directed above the volumetric flux vector. (Note that $\mathbf{k}$ is directed vertically downwards). Although these results were obtained by assuming (46), the resulting conditions for the allignment of $\mathbf{c}$ with either of $\mathbf{J}_M$, $\mathbf{J}_E$ or $\mathbf{J}_Q$ are dependent only on the thermodynamic properties of steam and water, and are independent of pressure gradients, permeability or porosity.

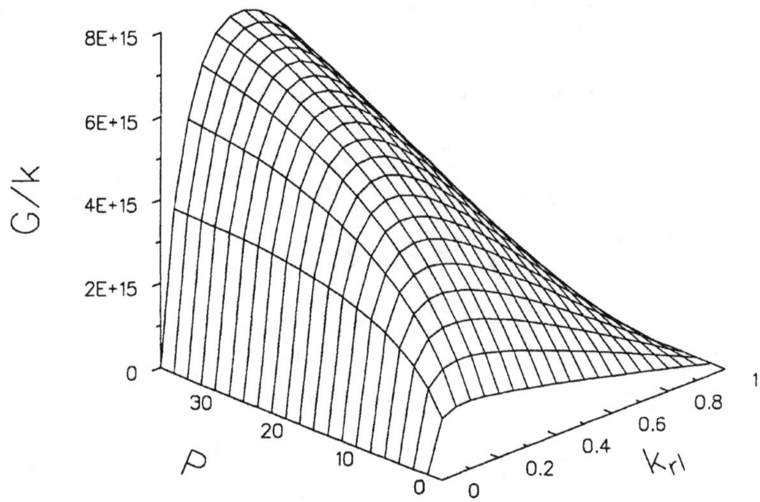

Figure 1: Three dimensional plot of G/k, in SI units, as a function of k_{rl} and pressure in bars.

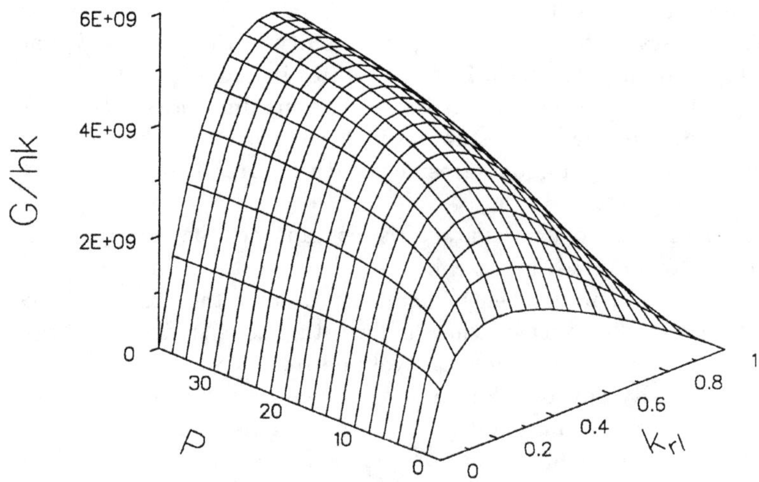

Figure 2: Three dimensional plot of G/kh, in SI units, as a function of k_{rl} and pressure in bars.

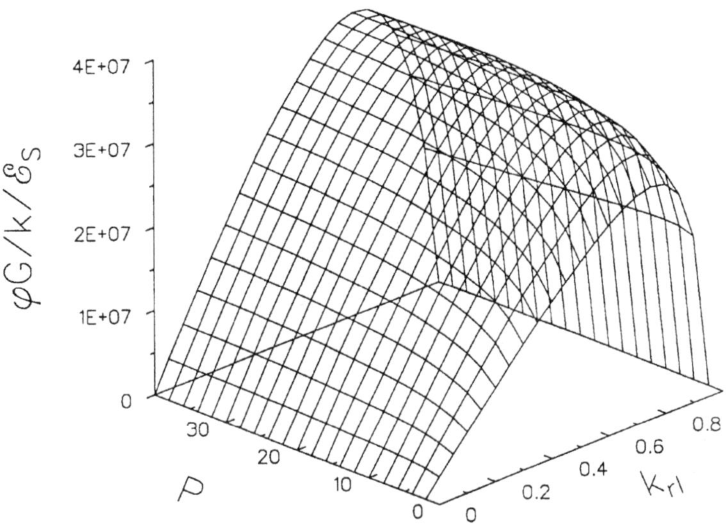

Figure 3: Three dimensional plot of $\phi G/k\mathcal{E}_S$, in SI units, as a function of k_{rl} and pressure in bars.

5. Discussion

This paper has presented a general discussion of saturation changes in a uniform porous media occupied by two phase fluid. In common with past work in geothermal reservoir engineering, capillarity pressure has been ignored. The two nonlinear parabolic equations of mass and energy conservation were rewritten as a single parabolic 'pressure equation', and a first order 'saturation equation'. This system of equations was termed singular, since the original system of equations which appear in conservation form to be fourth order, are then only third order.

Three mechanisms were identified as sufficient for describing saturation changes in a two phase region. Firstly, local boiling couples saturation changes to local changes in temperature (pressure) through essentially a latent heat relationship. Secondly, the variation of flowing enthalpy with position, in response to a mass flux, produces an energy imbalance, which can result in saturation changes. Thirdly, saturation changes can occur from a counterflow term, associated with equal and opposite vertical mass fluxes. Scalar functions were derived which were intimately related to these three mechanisms for changing saturation.

The characteristic velocity associated with the saturation equation was analysed, and shown to coincide with the direction of very weak shock propagation. This is of special interest, since there are several possible definitions of 'characteristic speeds'. (For example, consideration of the right hand side of (27) leads to another set of 'characteristics', in which the saturation derivatives in (28) are replaced by pressure derivatives). To terms first order in the change in saturation across a shock, the shock velocity direction bisects the angle between the two intersecting characteristics at a shock. Volumetric flux was shown to be continuous across a shock, where saturation and pressure gradients may be discontinuous. This allows a calculus of saturation derivatives to be formulated. One consequence was that saturation increases across a weak shock whenever the magnitude of the characteristic velocity increases with saturation.

New results were also presented on the geometry of flux vectors in a two phase region. For each spatial position, the three dimensional vectors $\mathbf{J}_M$, $\mathbf{J}_E$, $\mathbf{J}_Q$ and $\mathbf{c}$ were shown to always lie in the same vertical plane, and so such two phase regions are largely two dimensional, principally because the only forces applying result from gravity and the pressure gradient.

A vertical hierarchy of flux vectors was obtained, with members of the triple $\mathbf{J}_M$, $\mathbf{J}_E$, $\mathbf{J}_Q$ comprising mass, energy and volumetric flux vectors, never pointing below a preceding member. In a mixed flow region, characteristic velocity is nonzero, and tends to orientate vertically upwards with increasing saturation. In particular, it was shown that for a large class of relative permeability functions, the characteristics may point below the mass flux vector for small saturations, and may point above the volumetric flux vector for sufficiently large saturations.

Acknowledgement : The author has benefitted from a discussion with R. Young on the entropy condition.

References

Burnell, J.G., McNabb, A., Weir, G.J. and R. Young, 1989, Two phase boundary layer formation in a semi-infinite porous slab, *in press TIPM*

Burnell, J., Kissling, W., McGuinness, M., McNabb, A., Weir, G., White, S. and R. Young, 1990, Two-phase flow studies, submitted *1989 NZ Geothermal Workshop, Auckland University*

Garg, S.K. and Pritchett, J.W., 1977, On Pressure-Work, Viscous Dissipation and the Energy Balance Relation for Geothermal Reservoirs, *Advances in Water Resources* **1(1)** 41–47.

Grant, M.A., Donaldson, I.G. and Bixley, P.F., 1982 *Geothermal Reservoir Engineering* Academic Press, London.

McGuinness, M.J. and Pruess, K., 1987, Unstable heat pipes, in the *proceedings of the Ninth New Zealand Geothermal Workshop, Auckland University*

Transport in Porous Media **6**: 519–547, 1991

© 1991 *Kluwer Academic Publishers. Printed in the Netherlands.*

NUMERICAL SIMULATION AND HOMOGENIZATION OF TWO-PHASE FLOW IN HETEROGENEOUS POROUS MEDIA

BRAHIM AMAZIANE *, ALAIN BOURGEAT †, AND JOE KOEBBE ‡

Abstract. A mathematically rigorous method of homogenization is presented and used to analyze the equivalent behavior of transient flow of two incompressible fluids through heterogeneous media. Asymptotic expansions and H-convergence lead to the definition of a global or effective model of an equivalent homogeneous reservoir. Numerical computations to obtain the homogenized coefficients of the entire reservoir have been carried out via a finite element method. Numerical experiments involving the simulation of incompressible two-phase flow have been performed for each heterogeneous medium and for the homogenized medium as well as for other averaging methods. The results of the simulations are compared in terms of the transient saturation contours, production curves, and pressure distributions. Results obtained from the simulations with the homogenization method presented show good agreement with the heterogeneous simulations.

Key words. Heterogeneous porous media, two-phase flow, numerical simulation, finite element method.

1. Introduction. We have rigorously derived "homogenized" or "global" reservoir equations from exact local incompressible, immiscible, two-phase flow equations in a periodic heterogeneous medium for a single rock-type model including capillary and gravity effects, [2], [8], [9], [10], [11], and [12]. The homogenization of a miscible model without any dispersion term has been studied in [5], [6].

For the purposes here we assume that the local porosity $\Phi^\epsilon(x)$ and the local absolute permeability tensor $\mathbf{K}^\epsilon(x)$ are rapidly periodic oscillating functions in the space variable.

Then by homogenization methods [7], [26], and [29] i.e. a two-length scale asymptotic expansion

* UNIVERSITE DE PAU, CNRS - U.A. 1204 LABORATOIRE DE MATHEMATIQUES APPLIQUEES AV. DE L'UNIVERSITÉ - 64000 PAU - FRANCE

† UNIVERSITE DE SAINT-ETIENNE, EQUIPE D'ANALYSE NUMERIQUE U.A. 740, 23 RUE DU DR PAUL MICHELON - 42023 SAINT-ETIENNE CÉDEX 2 - FRANCE

‡ DEPARTMENT OF MATHEMATICS AND STATISTICS, UTAH STATE UNIVERSITY, LOGAN UTAH, 84322-3900 - USA

and H-convergence, we get "global" or "effective" permeabilities $\mathbf{K}^{\#}$ and porosity $\Phi^{\sim}$. In the multidimensional case, coefficients of the homogenized permeability tensor, $\mathbf{K}^{\#}$ are obtained by solving linear partial differential equations on the basic cell with periodic boundary conditions.

In this technique, the assumption of periodicity looks rather restrictive. But we must admit it is by now the only technique which allows us to address rigorously, nonlinear phenomena such as the two-phase flow system. We must also recall that this assumption of periodicity could be relaxed, since a non-uniform periodicity (i.e. coefficients depending on both the period ϵ and the space variable) is sufficient to get analogous results and to account for a wide range of physical heterogeneities.

We must also recognize that in the case of linear convection/diffusion equations, local stochastic heterogeneities assumptions give a result akin to this one obtained from the deterministic assumptions [27]. In that case, moreover, it could be proved that homogenization theory gives analogous results to these obtained by volume averaging [13]. In the same way, numerical results of multiscale techniques with periodicity assumptions are in good accordance, with the experimental data for various heterogeneous media [15].

2. The Two-Phase, Immiscible Model. Standing governing equations modeling the fluids in a porous reservoir $\Omega \subset R^n, 1 \leq n \leq 3$, over a time period $J =]0, T[$, [21], [28], are made up of equations describing conservation of mass in each phase coupled via a capillary pressure law $P_{cap}(\mathbf{x}, S)$, Figure (1), and relative permeability curves $k_{ri}(S)$ in each phase $i = w, o$. S_i and P_i are the saturation and pressure in each phase, where the subscripts w and o refer to water and oil, respectively.

The equations of conservation of mass and continuity for each phase are given by:

$$\Phi(\mathbf{x})\frac{\partial S_i}{\partial t} + \nabla \cdot \phi_i = 0 \quad x \in \Omega, t \in J \tag{1}$$

$$S_w + S_o = 1 \tag{2}$$

where ϕ_i is the velocity which is determined by the traditional extension of Darcy's law

$$\phi_i = -\frac{k_{ri}(S_i)}{\mu_i}\mathbf{K}(x)\nabla(P_i + P_g), \quad i = w, o \tag{3}$$

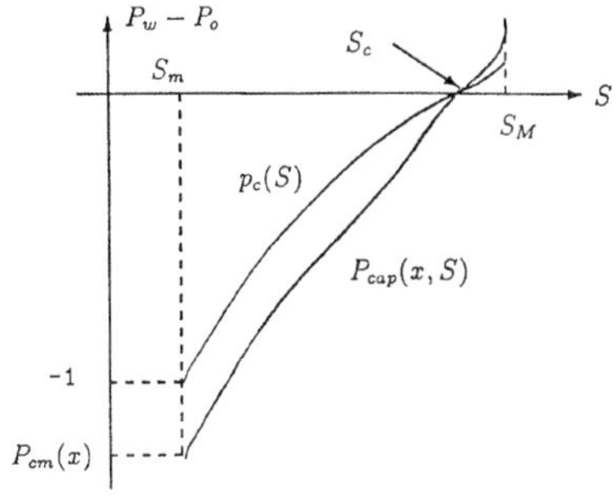

FIG. 1. *Capillary pressure curve at fixed* x, $P_w - P_o = P_{cap} = P_{cm}(x)p_c(S)$.

where μ_i is the viscosity of phase i, P_g the gravity potential, $\mathbf{K}(x)$ and $\Phi(x)$ the absolute permeability and porosity of the material.

The pressures in the two phases are related by the capillary pressure [19]

$$P_{cap}(x, S) = P_w - P_o = P_{cm}(x)p_c(S) \tag{4}$$

where $P_{cm}(x) > 0$ is the maximum of the absolute value of the capillary pressure at the point x and $p_c(S)$ is a dimensionless function such that $p_c'(S) > 0$, Figure (1).

The system (1)-(4) may be rearranged in a form which very closely resembles the system describing the displacement of two fully miscible fluids, [21]. This form is more suitable for mathematical purposes and will allow us to get results for both miscible and immiscible fluids as in [16], [24]. Define the total mobility, $\lambda(S)$, by

$$\lambda(S) = \frac{k_{rw}(S)}{\mu_w} + \frac{k_{ro}(S)}{\mu_o}$$

and the phase relative mobility as

$$\lambda_i(S) = \frac{k_{ri}(S)}{\mu_i \lambda(S)}.$$

The system is now rearranged using the "global" pressure variable [16], [24], given by

$$P = \frac{1}{2}(P_w + P_o) + \frac{1}{2}P_{cm}(x)\gamma(S) \tag{5}$$

where

$$\gamma(S) = \int_0^{p_c} (\lambda_w - \lambda_o)[p_c^{-1}(\tau)]d\tau.$$

Combining equations (1)-(4) we obtain

$$\mathbf{q}_t = -\lambda(S)\mathbf{K}(x)\nabla P + \lambda(S)[\gamma_1(S)\mathbf{q}_1 + \gamma_2(S)\mathbf{q}_2] \tag{6}$$

$$\nabla \cdot \mathbf{q}_t = 0 \tag{7}$$

with $\mathbf{q}_t = \phi_w + \phi_o$ is defined as the total fluid velocity and

$$\Phi(x)\frac{\partial S}{\partial t} + \nabla \cdot (\mathbf{R} + \lambda_w(S)\mathbf{q}_t + b_1(S)\mathbf{q}_1 + b_2(S)\mathbf{q}_2) = 0 \tag{8}$$

where $\mathbf{R} = -P_{cm}(x)\mathbf{K}(x)a(S)\nabla S$ is a diffusion term due to the capillary pressure. $\mathbf{q}_1 = -\mathbf{K}(x)\nabla P_{cm}(x)$ is a convective term due to spatial variation in the capillary pressure, and $\mathbf{q}_2 = -\mathbf{K}(x)\nabla P_g(x)$ is a convective term due to the gravity force. Also

$$\begin{aligned} a(S) &= \lambda(S)\lambda_w(S)\lambda_o(S)p_c'(S) \geq 0, \quad a(0) = a(1) = 0 \\ \gamma_1(S) &= \int_{S_c}^{S} \lambda_w'(\tau)p_c(\tau)d\tau \\ \gamma_2(S) &= \lambda(S)\frac{\rho_w\lambda_w(S) + \rho_o\lambda_o(S)}{\rho_m} \end{aligned}$$

with ρ_i, the phase density of the ith fluid and $\rho_m = \frac{1}{2}(\rho_w + \rho_o)$ the mean density.

Finally, define

$$\begin{aligned} b_1(S) &= \lambda(S)\lambda_w(S)\lambda_o(S)p_c(S) \\ b_2(S) &= \lambda(S)\lambda_w(S)\lambda_o(S)\frac{\rho_w - \rho_o}{\rho_m} \end{aligned}$$

Note that Equation (6) is frequently called the "global" pressure equation. It is a family of elliptic equations (one for each $t \in J$) with a saturation dependent coefficient.

Equation (8) may be considered a saturation (or concentration) equation because it is usually a convection-dominated parabolic equation for saturation (or concentration) with coefficients depending on pressure through the total Darcy velocity. In the case of immiscible displacement, the diffusion term $\mathbf{R}$ in the saturation equation vanishes at some points ($S = 0$ or $S = 1$). The equation degenerates at these points.

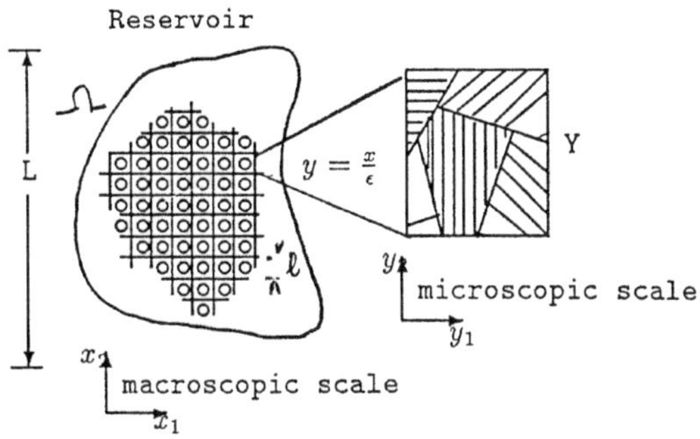

FIG. 2. *Illustration of the macroscopic scale and a microscopic view of a cell.*

3. General Presentation of Homogenization. Let u be a quantity describing a process on two different scales. This could happen for instance with any process (heat conduction, fluid flow, plate bending, etc,...) taking place in a heterogeneous region Ω. In this case, there is a small scale which is the scale of the heterogeneities l and a large scale which is the scale of the region Ω denoted by L. The quantity u will depend strongly on these two length scales and more precisely on their ratio $\epsilon = \frac{l}{L}$.

The purpose of homogenization method is to, by means of mathematical analysis, find the homogeneous properties of the region Ω which give a "globally equivalent" process through a homogeneous medium. This "globally equivalent" medium will be called "homogenized" and the parameters characterizing this medium will also be called "homogenized" or "effective" . The "homogenized" process will be described by the homogenized quantity u_0, [7], [26], [29].

From a mathematical point of view. to homogenize is simply to find the equation satisfied by u_ϵ as ϵ tends to zero, and thus find u_0, the limit of u_ϵ.

It should be noticed, however, that the "homogenized" equation could be different from the "heterogeneous" equation. For instance, the Stoke's equation in a porous medium (heterogeneous domain Ω made of obstacles and canals) gives a "homogenized" equation which is the Darcy law. Even in the simplest case, the "effective" behavior is never straightforward [30].

To understand the difficulties we may look at the stationary heat equations

$$\nabla \cdot (\mathbf{A}^\epsilon(x)\nabla u_\epsilon) \quad = \quad f$$

(9)
$$\mathbf{A}^{\epsilon}(x) = \mathbf{A}(x, \frac{x}{\epsilon})$$

where $\mathbf{A}(x, \frac{x}{\epsilon})$ is a rapidly oscillating function. Although the effective (or homogenized) equation is of the same type,

(10)
$$\nabla \cdot (\mathbf{A}^{\#}(x) \nabla u_0) = f$$

where $\mathbf{A}^{\#}(x)$ no longer depends on ϵ. It is proven that $\mathbf{A}^{\#}(x)$, except in some very special cases, is neither the arithmetic mean nor the harmonic mean of $\mathbf{A}(x, \frac{x}{\epsilon})$. This is due to the fact that $\mathbf{A}(x, \frac{x}{\epsilon})$ and ∇u_ϵ are two oscillatory quantities. Their product does not oscillating around the product of their arithmetic means. This can be seen in the example of $V(x, \epsilon) = sin(\frac{2\pi x}{\epsilon})$ oscillating around its mean of zero and $W(x, \epsilon) = V(x, \epsilon) \times V(x, \epsilon)$ which does not oscillate around the product of the arithmetic means (which would be zero) but oscillates around the arithmetic mean of W which is $\frac{1}{2}$.

These types of problems have been addressed by various authors in the field of Petroleum Engineering. Most of the work is based on heuristic considerations and takes into account only incompressible one-phase flow in heterogeneous porous media. Moreover, the most common conclusion of these works is that the "effective" permeability lies between a harmonic and arithmetic average, [14], [25], or is given by the geometric mean [31].

However, because "homogenization" is a rigorous mathematical method, some very precise hypotheses will be imposed. The first hypothesis is that we should know the mean around which coefficients are oscillating which in some sense is a "weak" limit of the coefficient. The easiest way to have such an assumption is to assume that the coefficients are rapidly and quasi-periodically oscillating, i.e. depend on x and $\frac{x}{\epsilon}$, with period ϵ. Any other assumption that gives the "weak" limits of the coefficients could have been used. The second hypothesis is that the region under consideration does not intersect any boundaries where boundary layers occur.

4. Theoretical Homogenization of Two-Phase Flow. In the framework described below we have considered two cases. The first one is the immiscible flow of two incompressible fluids, including the effects of capillary pressure and gravity and the second one is the case of miscible flow of two incompressible fluids including the effects of dispersion and gravity. In both cases we start from a description of the flow at the level of the heterogeneous porous medium (the small scale l is the scale of the heterogeneities of the porous medium) and we seek a homogenized description at the macroscopic level of the reservoir (of length scale L), Figure (2).

To get rigorous results we have used the same type of assumptions as above, and some additional assumptions essentially due to the difficulty of these problems. In the case of a miscible displacement, we assume that the dispersion tensor $\mathbf{D}$, depends only on the space variable x and on the concentration C. Moreover it is assumed at a fixed C that this tensor is a rapidly periodically oscillating function, taking different values in each type of porous media $\mathbf{D} = \mathbf{D}_1(x) \times \mathbf{D}_2(C)$. The porosity Φ and absolute permeability tensor $\mathbf{K}$ of the rock take on different values in each type of porous media and then are rapidly quasi-periodically oscillating functions, $\Phi^\epsilon(x) = \Phi(x, \frac{x}{\epsilon})$, $\mathbf{K}^\epsilon(x) = \mathbf{K}(x, \frac{x}{\epsilon})$, and $\mathbf{D}_1^\epsilon(x) = \mathbf{D}_1(x, \frac{x}{\epsilon})$.

In the case of an immiscible displacement we assume that the capillary pressure curve $P_{cap}(x, S) = P_{cm}(x) \times p_c(S)$, does not depend on the heterogeneities and then is assumed to depend only on the space variable x and the saturation.

In the first part we are assuming that the relative permeability and capillary pressure, $k_{ri}(S)$ and $p_c(S)$, have the same shape all over the domain of the reservoir Ω. Hence there is no jump in the saturation at the interfaces between heterogeneities. This would be rephrased by saying that we considered a field containing a single rock type. Under these assumptions we get the results below.

4.1. Theoretical Results on a Single Type Rock Field. From [2], [10], and [12] in the case of the periodicity assumption the main results are

1. *Equations describing the homogenized flows are of the same type that the equations of the flow through the heterogeneous medium. There is an effective porosity $\Phi^\sim$, effective permeabilities $K^\#$, (and an effective dispersion tensor $\mathbf{D}_1^\#$) which do not depend on x or $\frac{x}{\epsilon}$.*
2. *$\Phi^\sim$ is the arithmetic average.*
3. *$\mathbf{K}^\#$ (and $\mathbf{D}_1^\#$) are given by the same method as in the linear case, i.e. for instance $\mathbf{K}_{i,j}^\#$ is the arithmetic average of the sum of the coefficients $\mathbf{K}_{i,j}$ and of a complimentary term $\alpha_{i,j}$.*

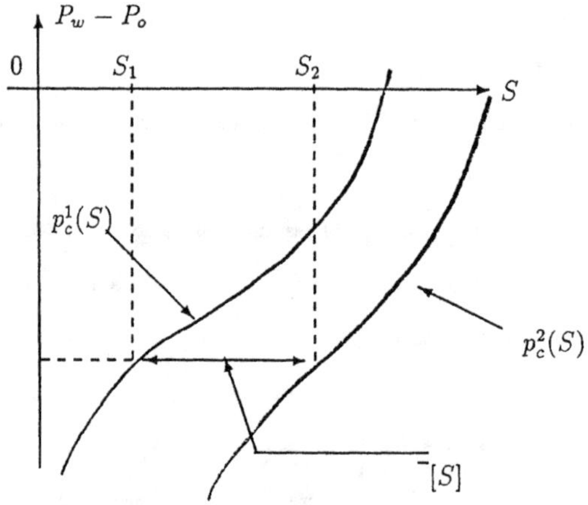

FIG. 3. *An example of the capillary pressure curves for the two media.*

This complementary term $\alpha_{i,j}$ is given by

$$\alpha_{i,j} = \sum_{k=1}^{n} K_{i,k} \frac{\partial w_j(y)}{\partial y_k} \qquad 1 \le i, j \le n$$

where $w_j(y)$ is the periodic solution of a linear PDE in the variable $y = \frac{x}{\epsilon}$. See section 5.

Remark 1: The results given above allow us to compute the homogenized flow in all cases of heterogeneities such as anisotropic media, various shapes of heterogeneities and several types of heterogeneities.

Remark 2: These results are consistent with the widely used rules of averaging when the flow is orthogonal or parallel to the stratas in a stratified porous medium.

Remark 3: In general cases, even with each type of porous media being isotropic, we have homogenized effective tensors which are significantly anisotropic.

4.2. Theoretical Results on a Many Type Rock Field. We may relax the assumption on the capillary pressure and the relative permeability curves, that is allow those curves to change in each type of porous media. In each type j, there is a p_c^j, Figure (4), and k_{ri}^j functions and hence there will be a saturation jump at the interface between each type of media [19].

For sake of simplicity, we will consider the case of a horizontal field Ω which contains only two different types of rock ($j = 1, 2$) and we will assume $P_{cm}(x) = C^{te}$. Note that in the sequel, the subscript j will refer to the jth type of rock.

Let Ω_j, $j = 1, 2$ be the spatial domain of type j and Γ_{12} the interface between Ω_1 and Ω_2. There are in each Ω_j, $j = 1, 2$ the same equations as (6)-(8), Section 2.

$$\mathbf{q}_t^j = -\lambda^j(S_j)\mathbf{K}(x)\nabla P_j \tag{11}$$

$$\nabla \cdot \mathbf{q}_t^j = 0 \tag{12}$$

in $\Omega_j \times J$ and

$$\Phi(x)\frac{\partial S_j}{\partial t} + \nabla \cdot \mathbf{V}_w^j = 0 \tag{13}$$

in $\Omega_j \times J$, where

$$\mathbf{V}_w^j = \mathbf{R}^j + \lambda_w^j(S_j)\mathbf{q}_t^j \tag{14}$$

and at the interface, the continuity of the physical quantities, P_w and P_o (pressures) and the flux of each of the two fluids, gives the interface conditions:

1. Continuity of the total flux:

$$[[\mathbf{q}_t \cdot \nu]] = 0 \tag{15}$$

2. Continuity of the water flux:

$$[[\mathbf{V}_w \cdot \nu]] = 0 \tag{16}$$

3. Continuity of the fluid pressure gives a jump on the "global" pressure:

$$[[P]] = [[\gamma(S)]] \tag{17}$$

4. Continuity of the capillary pressure gives a jump on the saturation S, Figure (3)

$$p_c^1(S_1) = p_c^2(S_2) \tag{18}$$

where ν is the outward normal to Γ_{12} and the double brackets mean, as usual, the jump across Γ_{12}.

Essential difficulties of this problem are the nonlinearity arising in the interface condition (18), the discontinuity of the saturation and of the "global" pressure given by (18) and (17) on the interface and the supplementary coupling between the saturation equation (13) and the "global" pressure equation (11).

In a first step, we have linearized the equations and the capillary pressure curves, decoupled the system (cross mobility curves), assumed the global pressure jump at the interface is of order $O(\epsilon)$. By taking $P_{cap}(S)$ as an auxiliary variable we find homogenized equations are of the same type as the original. But the absolute permeability and porosity are given by the other expressions than the standard ones in Section 3. [2], [3].

For the more general case with the coupled system (11)-(18) we introduce again the auxiliary variable $P_{cap}(S)$ (which is now a nonlinear function of S). This leads to a variational formulation associated with the semi-discretization in time. Then a mixed form is used for the "global" pressure.

Up to now we have a mathematical proof of existence of the solution of this system. the homogenization of this problem is currently underway. It appears that the homogenized system will not be of the same form as the initial one except in special cases.

5. Numerical Simulations.

5.1. Numerical Computations of the Homogenized Tensor. In general n-dimensional cases, to obtain the homogenized tensor, we have to solve n so-called "local" problems which are linear PDEs on a basic cell with periodic boundary conditions. The idea is to obtain the function $w_j(y)$, for $j = 1 \dots n$ (see Section 4). Namely,

$$-\nabla \cdot (\mathbf{K}(y)\nabla w_j(y)) = \nabla \cdot (\mathbf{K}(y)e_j) \tag{19}$$

in Y and $w_j(y)$, Y-periodic, where e_j is the jth standard basis vector.

To compute the solution of (19) by a finite element method, we write down the variational formulation of this problem,

$$\int_Y \mathbf{K}(y)\nabla w_j \cdot \nabla v = \int_Y \nabla \cdot (\mathbf{K}(y)e_j)v \tag{20}$$

for all v belonging to the Sobolev space $H^1(Y)$, and Y-periodic. A numerical code [4] has been written to solve problem (20) and compute the homogenized tensor obtained in Section 4.1, for any heterogeneous distributions in a given cell. The results of the approximate solution of Equation (20) are used below in several test problems.

5.2. Numerical Simulations of Two-Phase Flow in a Reservoir. In order to test the validity of the theoretical results, we have run various simulations for incompressible two phase flow. Comparisons have been done between heterogeneous and homogenized simulations for saturations and pressures, visualized by means of contour plots, cross-sections of surfaces, and through the production curves associated with the amount of resident fluid (oil) produced in the simulation. There are essentially two groups of simulations that have been performed.

1. Simulation on core samples and stratified quarter five-spot patterns have been performed using the BIDIMIX code (INRIA-ELF-IFP, [17], [18], [19], [20]) at INRIA in Rocquencourt France. These simulations include the testing of the influence of boundary layers and the ratio of the scale $\epsilon = \frac{l}{L}$ on the accuracy of the homogenized model. Various Peclet numbers and mobilities have been tested (see [2],[3]).
2. 2-D heterogeneities in a quarter five-spot pattern have been tested at the University of Wyoming in Laramie Wyoming and Utah State University in Logan Utah. The goal of these studies is to test by numerical simulation, homogenized coefficients predicted by the theory above. As a means of comparison the arithmetic mean, harmonic average, and geometric mean have been used to compute other "effective " coefficients.

The following graphs show some of the results from the simulations done at the University of Wyoming and Utah State University using the SIM2P2D code [1], [22], and [23].

5.3. Presentation of Numerical Results. Numerical experiments to test the homogenization method described here have been performed on a variety of problems including two-phase miscible and immiscible displacement regimes. Also a variety of flow parameters including the viscosity ratio between the two fluids and the permeability ratio between different regions in the porous medium. The results will be presented below in four parts delineating four representative test problems. In

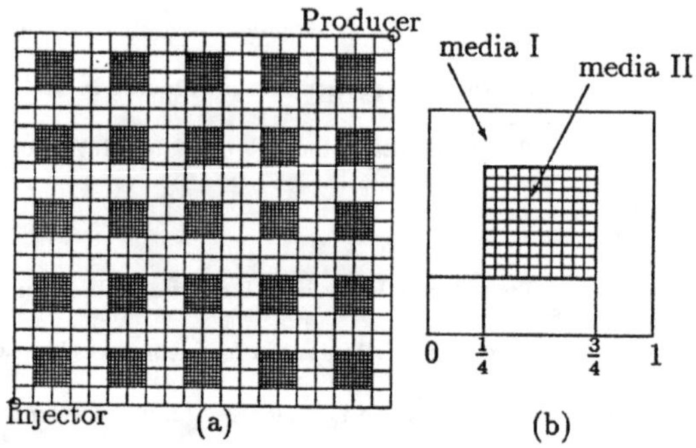

FIG. 4. *The general test pattern is shown in (a) with the grid blocks overlayed. The close up view of the cells is shown in (b). The computational grid is shown in (a) in the region where media I resides. The grid in media II in (b) is not the computational grid*

all test problems below a quarter five-spot pattern was used on a unit square. The simulator used in obtaining the numerical results features the use of an IMPES solution of the pressure/velocity equation and saturation equation. The pressure and velocity (determined by Darcy's law) equations are solved implicitly in the simulator to obtain approximations to the pressure and velocity at a new time level and then the saturation of the water phase is obtained via some explicit update given the total fluid velocity [22], [23]. The explicit update of the saturation equation for the test problems below was performed using one-point upwinding.

The simulations done in all cases had a fixed porosity, Φ, fixed injection and production rates at the wells, and the fluids were assumed to be incompressible. Below it will be seen that for incompressible fluids with fixed injection and production that the saturation contours are not very sensitive to a change in the "effective" coefficients. However the pressure distribution in the domain will be quite different to compensate for the difference in the coefficients.

The type of displacement in the formulation of the equations is described by the types of relative permeability curves describing the ability of the phases to flow in the medium. The test problems all used the same relative permeability curves for the two phases. The relative permeabilities used were

$$k_{rw}(S) = S^2(3-2S) \qquad k_{ro}(S) = 1 - k_{rw}(S).$$

Finally, it should be noted that the permeability tensor in all cases below was taken to be diagonal

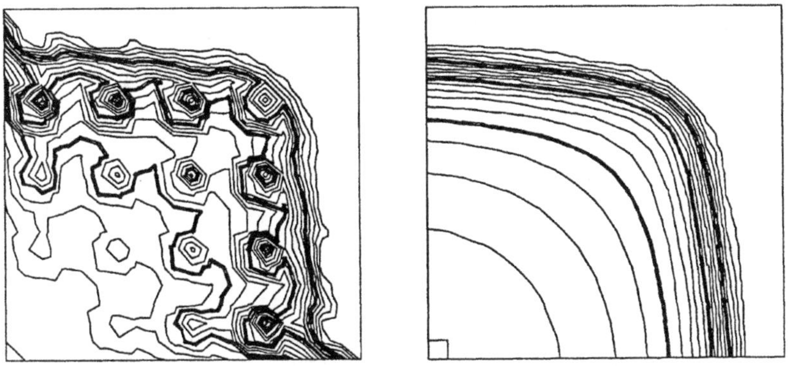

(a) (b)

FIG. 5. *The saturations contours before breakthrough at the production well for (a) the heterogeneous simulation and (b) for the homogenized simulation with a* **5x5 cell** *pattern for test problem #1.*

with components K^j_{xx} and K^j_{yy}, where j is an index that refers to different regions in the domain. If a cell has two regions of permeability then j would be defined for region I and II (see Figure (4)).

5.3.1. Test Problem #1. The first test problem involved simulation of a quarter five-spot pattern with heterogeneities as shown in Figure (4). Figure (4a) shows a typical domain for the simulation where a repeated pattern of heterogeneous cells occurs. Figure (4b) shows a close up view of one of the heterogeneous cells. The shaded region in the interior of the cell is of low permeability, while the border is a region of higher permeability. For the simulation done for test problem #1 the relative dimensions for the two regions of differing permeabilities is shown in Figure (4b). Typical values chosen through out the numerical experiments for the ratio of the permeabilities in the two regions are 10:1, 100:1, and 1000:1. In Figure (4) the computational grid used in the simulations is shown in the depiction of the entire domain. The shading in Figure (4b) should not be confused with the computational grid. This shading appears only to show a difference between the two regions. The "darker" region is the region of lower permeability

For test problem #1 the permeability ratio was taken as 10:1. That is the two diagonal components, K^j_{xx} and K^j_{yy} are defined by $K^I_{xx} = K^I_{yy} = 10md$ and $K^{II}_{xx} = K^{II}_{yy} = 1md$. In this case the homogenized value of the permeability is given by $K^{\#}_{xx} = K^{\#}_{yy} = 6.52$md, while for the arithmetic mean, $K^a_{xx} =$

(a) (b)

FIG. 6. *The saturations contours after breakthrough at the production well for (a) the heterogeneous simulation and (b) for the homogenized simulation with a* 5x5 cell *pattern for test problem #1.*

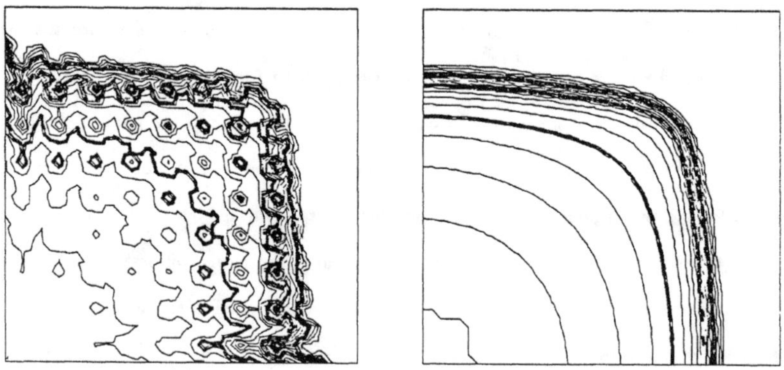

(a) (b)

FIG. 7. *The saturations contours before breakthrough at the production well for (a) the heterogeneous simulation and (b) for the homogenized simulation with a* 10x10 cell *pattern for test problem #1.*

(a) (b)

FIG. 8. *The saturations contours after breakthrough at the production well for (a) the heterogeneous simulation and (b) for the homogenized simulation with a* **10x10 cell** *pattern for test problem #1.*

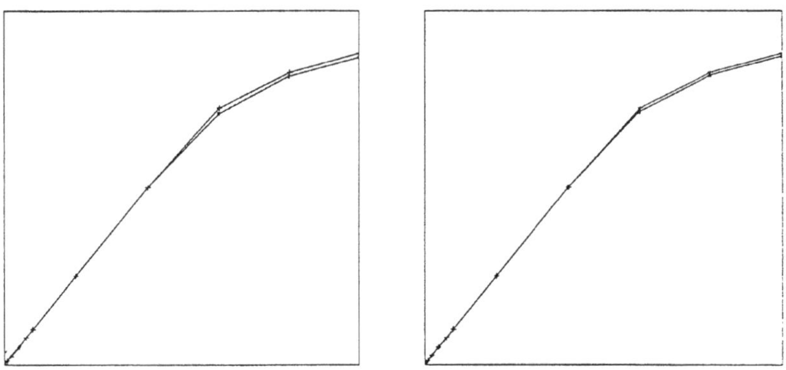

(a) (b)

FIG. 9. *The production curves for the heterogeneous and homogenized simulations are plotted for (a) a* **5x5 cell** *pattern and (b) a* **10x10 cell** *pattern for test problem #1.*

simulation performed involved a 5x5 pattern of cells as shown in Figure (4a) (20x20 computational grid) and the second simulation involved a 10x10 pattern of cells (40x40 computational grid) on the same domain.

Figure (5a) shows the saturation contours for the 5x5 cell pattern before breakthrough occurs at the production well. Figure (5b) shows the saturation contours at the same time on the homogenized medium. Figures (6a) and (6b) show the saturation contours after breakthrough has occured at the production well. Figures (7) and (8) show analogous results in the 10x10 cell pattern for the second simulation in test problem #1. Figure (9) shows the production curves measuring the amount of oil produced at the production well for the two simulations done under this test problem.

The results indicate reasonable agreement between the heterogeneous and homogenized simulations. The production curves indicate that as the cell is made smaller with respect to the domain length scale ($\epsilon \rightarrow 0$) the approximation is better as one would expect.

5.3.2. Test Problem #2. Under Test problem #2, two simulations were again done. The simulations involved the same cell patterns as in test problem #1. In this case, the viscosity ratio of the fluids was chosen to be 50:1. That is the resident oil phase was 50 time as viscous as the water phase being injected. Note that the homogenized value for the permeability will remain the same in this test problem.

The results for the second test problem are shown in Figures (10) through (14). The graphics from the simulations are shown in the same order as in test problem #1. The results of the simulations again show good agreement between the heterogeneous simulations and the homogenized simulations. Again the production curves show similar improvement as the size of the cells is decreased relative to the length scale of the whole domain as in test problem #1.

5.3.3. Test Problem #3. The third test problem uses the same pattern of heterogeneitics as the first two test problems, but changes the permeability ratio to 100:1. The associated homogenized value of the permeability is $K^{\#}_{xx} = K^{\#}_{yy} = 59.2$. The viscosity ratio is 50:1 as in test problem #2. It should be noted at this point that the viscosity ratio being 50:1 only corresponds to a mobility ratio of about 0.40 at the front for the cubic relative permeability curves used in all the problems. Thus the

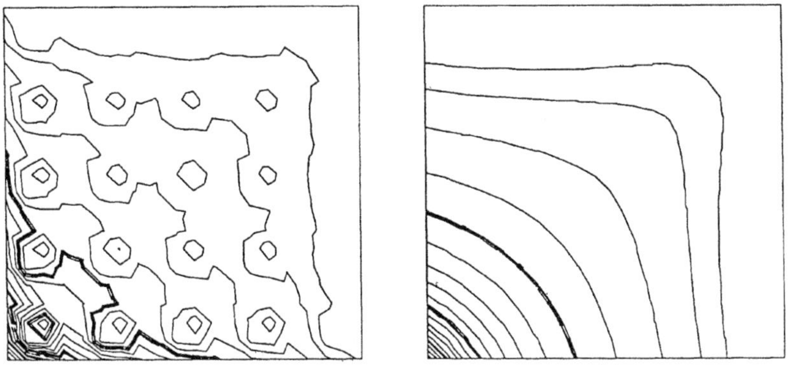

(a) (b)

FIG. 10. *The saturations contours before breakthrough at the production well for (a) the heterogeneous simulation and (b) for the homogenized simulation with a* 5x5 cell *pattern for test problem #2.*

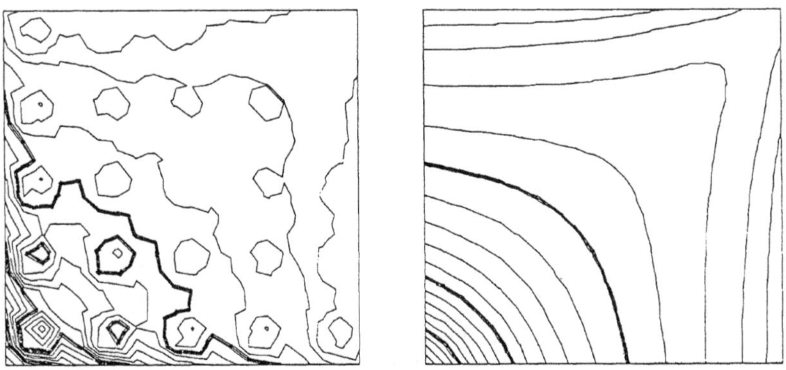

(a) (b)

FIG. 11. *The saturations contours after breakthrough at the production well for (a) the heterogeneous simulation and (b) for the homogenized simulation with a* 5x5 cell *pattern for test problem #2.*

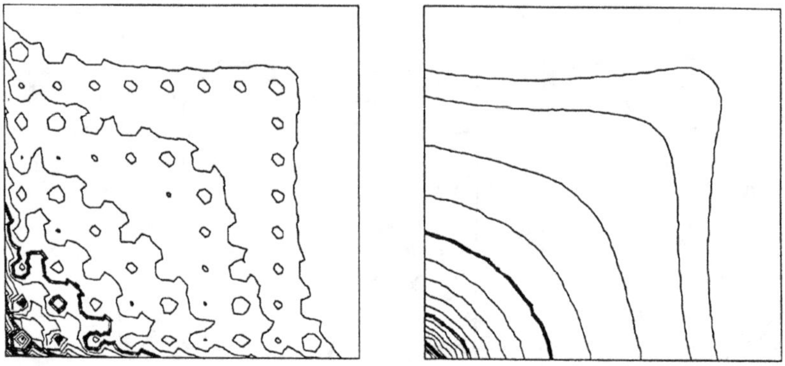

(a) (b)

FIG. 12. *The saturations contours before breakthrough at the production well for (a) the heterogeneous simulation and (b) for the homogenized simulation with a* 10x10 cell *pattern for test problem #2.*

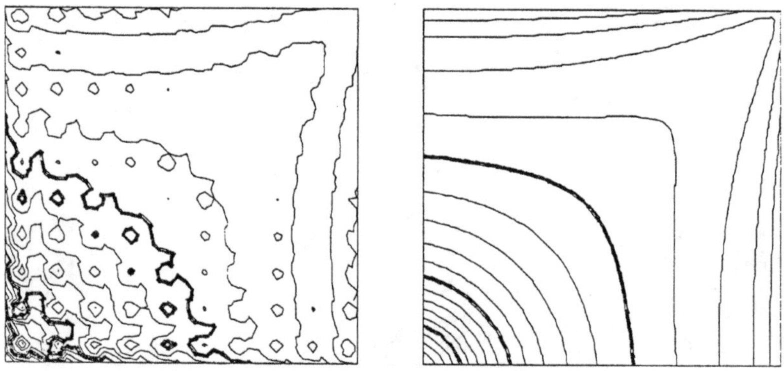

(a) (b)

FIG. 13. *The saturations contours after breakthrough at the production well for (a) the heterogeneous simulation and (b) for the homogenized simulation with a* 10x10 cell *pattern for test problem #2.*

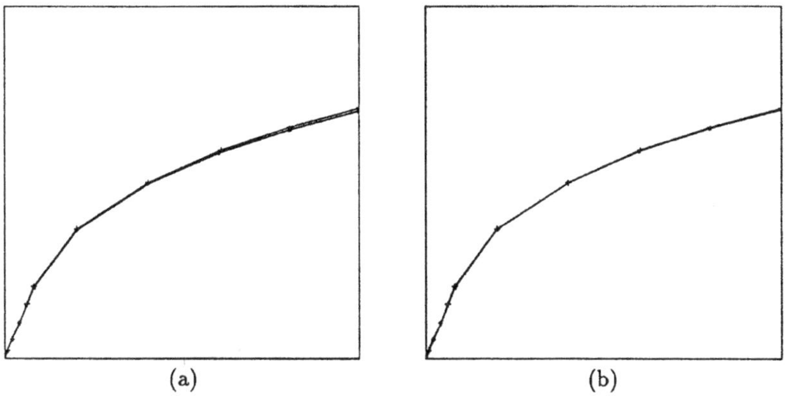

FIG. 14. *The production curves for the heterogeneous and homogenized simulations are plotted for (a) a* **5x5 cell** *pattern and (b) a* **10x10 cell** *pattern for test problem #2.*

should be noted at this point that the viscosity ratio being 50:1 only corresponds to a mobility ratio of about 0.40 at the front for the cubic relative permeability curves used in all the problems. Thus the fingering effect will not be as great for this problem. For this test problem only one set of simulations is presented. The case of a 5x5 pattern of cells is considered in the figures.

Again the results in the first two figures indicate reasonable agreement between the saturation contours both before and after breakthrough. However the production curves shown in Figure (17) show a large discrepancy in the results of the simulations on the heterogeneous medium and the homogenized medium. The reader should note that the theory is a perturbation method in the "small" parameter ϵ. In this case both permeability ratio being large and the fact that the value of ϵ is large relative to the domain length scale combine to cause the inconsistency in the results. If the number of cells becomes large ($\epsilon \rightarrow 0$), one could expect the homogenization theory described here to do a better job. We should note that in this case the values for the homogenized and arithmetic mean coefficient values are close.

5.3.4. Test Problem #4. In test problem #4, the pattern was changed slightly and is shown in Figure (18). In this case the cell pattern is changed only in the dimensions of the regions containing the two media. The width of the high permeability region is only one tenth the total cell width limiting

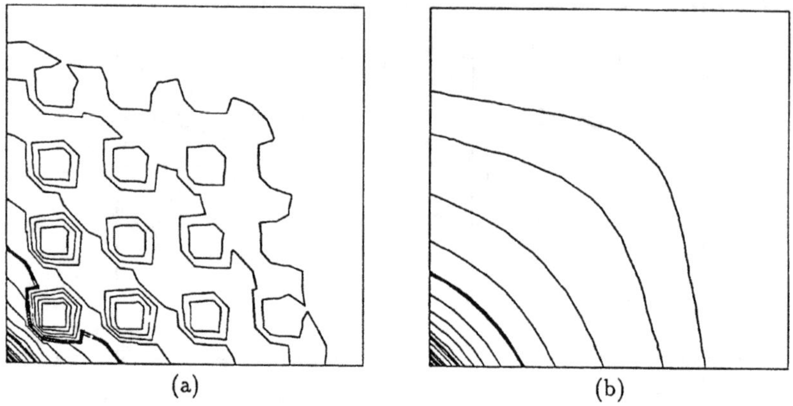

FIG. 15. *The saturations contours before breakthrough at the production well for (a) the heterogeneous simulation and (b) for the homogenized simulation with a* 5x5 cell *pattern for test problem #3.*

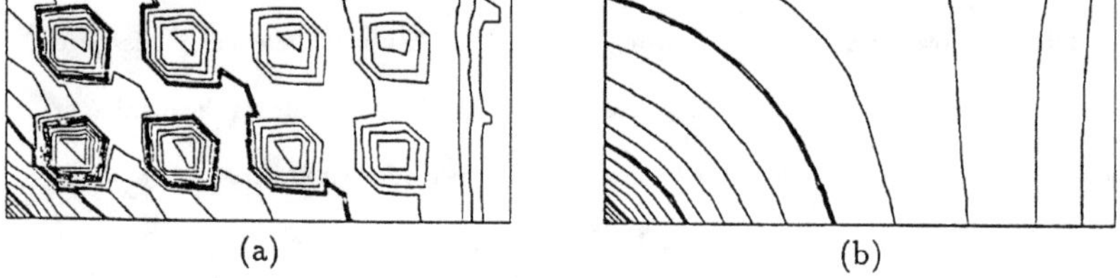

FIG. 16. *The saturations contours after breakthrough at the production well for (a) the heterogeneous simulation and (b) for the homogenized simulation with a* 5x5 cell *pattern for test problem #3.*

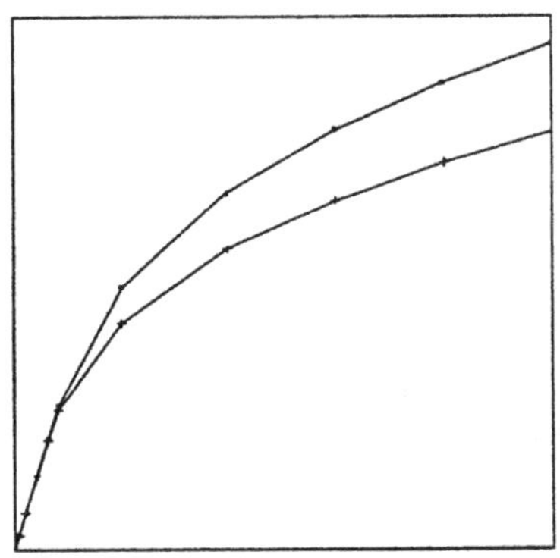

FIG. 17. *The production curves for the heterogeneous and homogenized simulation are plotted for a* 5x5 cell *pattern for test problem #3.*

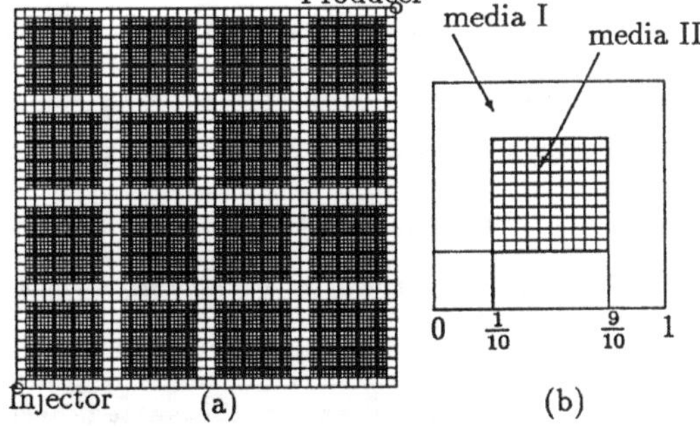

FIG. 18. *The general test pattern is shown in (a) with the grid blocks overlayed. The close up view of the cells is shown in (b). The computational grid is shown in (a) in the region where media I resides. The grid in media II in (b) is not the computational grid*

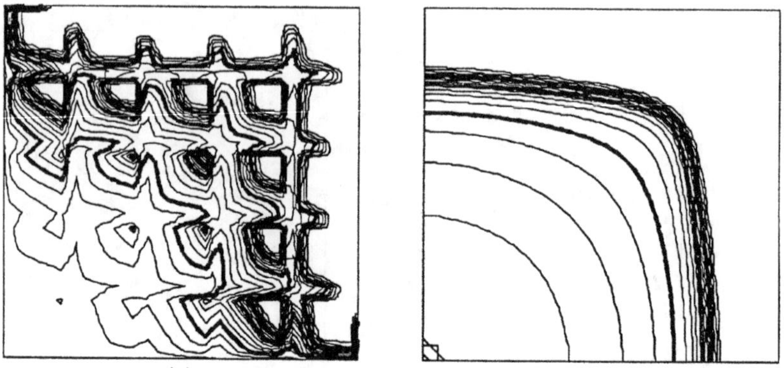

(a) (b)

FIG. 19. *The saturations contours before breakthrough at the production well for (a) the heterogeneous simulation and (b) for the homogenized simulation with a* 5x5 cell *pattern for test problem #4.*

the flow path relative to the previous test problems. The viscosity ratio was fixed at 1:1 and the permeability ratio is fixed at 10:1 as in the first test problem. The homogenized permeability value is $K^{\#}_{xx} = K^{\#}_{yy} = 3.106$. The results below were compared to simulations performed using "effective" permeabilities obtained by finding the arithmetic mean, $K^{a}_{xx} = K^{a}_{yy} = 4.24$, the harmonic average, $K^{h}_{xx} = K^{h}_{yy} = 1.48$, and the geometric mean, $K^{g}_{xx} = K^{g}_{yy} = 3.33$. We should note that in this case the values for the homogenized and geometric mean coefficient values are very close.

The results shown in Figures (19) through (21) show the same type of agreement as in the first two test problem results presented. The saturation curves and production curves do not show any great differences in the homogenized or arithmetic mean simulations. To see significant differences we turn to the pressure distributions for the different methods for determining the effective parameters.

Figures (22) through (25) show the pressure contours as well as the cross-sections of the pressures between the wells for the four methods. The contours do not show up a great deal of difference between the methods, but the cross-sections show that there is a significant difference in the approximation to the pressure variable. The homogenized and geometric mean effective parameters seem to provide the best results in determining an approximation to the pressure, but the fact that these results are close is due to the particular test problem. In test problem #1, the arithmetic mean was closer to the homogenized value, while the geometric mean value was quite different. The relative closeness of the

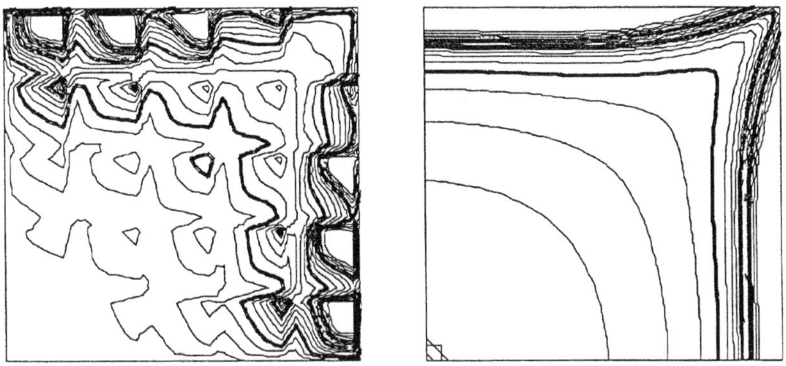

(a) (b)

FIG. 20. *The saturations contours after breakthrough at the production well for (a) the heterogeneous simulation and (b) for the homogenized simulation with a* 5x5 cell *pattern for test problem #4.*

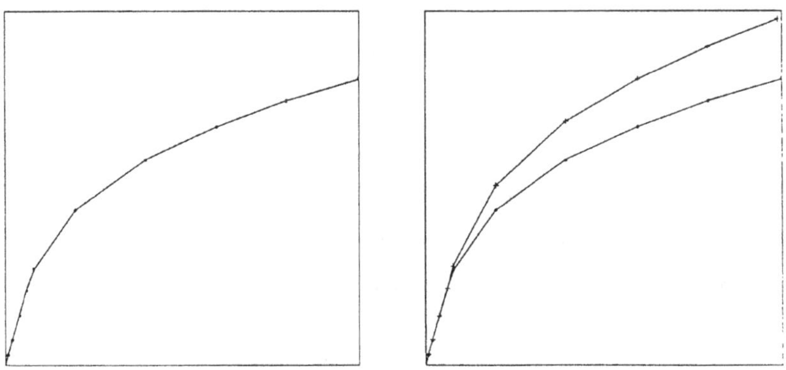

FIG. 21. *The production curves for the heterogeneous and homogenized simulations are plotted for a* 5x5 cell *pattern with (a) the homogeinzation proposed here and (b) using the* arithmetic *mean with a* 5x5 cell *pattern for test problem #4.*

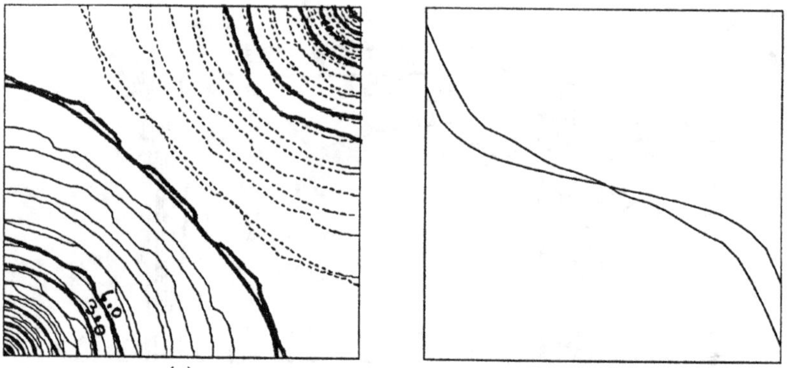

(a) (b)

FIG. 22. *The pressure contours for the heterogeneous and* **arithmetic** *average simulations are shown in (a) and the comparison of the cross-sections between wells is shown in (b) for a* **5x5 cell** *pattern on test problem #4.*

values obtained by the other methods is by chance and not due to any rigorous theory.

As a final comment on the comparison of the results from the various methods of averaging to obtain "effective" permeabilities, it should be noted that if a finite difference scheme is used that obtains the pressures directly one can expect that the numerical differentiation used to obtain the gradient of the velocity for use in the saturation equation will in general be inaccurate. In the simulations done in these numerical experiments, a mixed finite element method was used which obtains accurate velocities simultaneously with the approximation of the pressure variable. Very little change was noticed in the velocities in the numerical experiments for various permeabilities. However, the pressure may be necessary in a simulation whether for use in a numerical differentiation or in some other capacity. For example, if a five-point finite difference scheme is used to approximate the pressure variable and then the fluid velocity is computed using a finite difference form of Darcy's law, the results could be corrupted due to the differencing of the approximate pressures. Using inaccurate permeabilities with inaccurate pressure gradiants in turn would corrupt the transport of the two-phases in the simulation. One would expect that the saturation contours would be quite different in this case. Thus it is important to obtain the best possible choice for the "effective" coefficients. The mathematical theory and numerical experiments suggest that the best choice is achieved by using the homogenization technique described here.

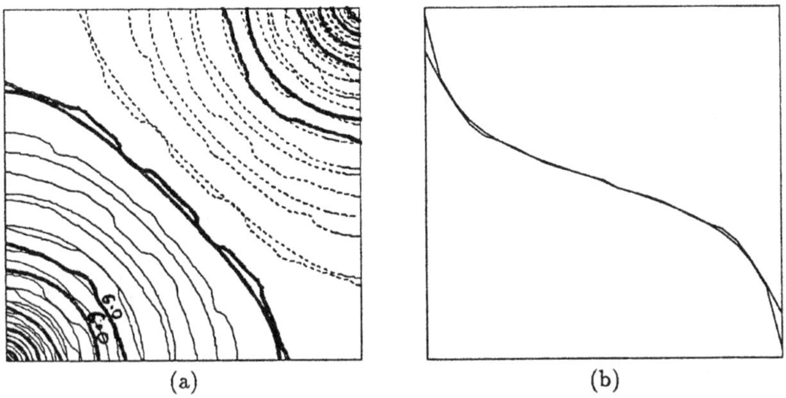

FIG. 23. *The pressure contours for the heterogeneous and* **homogenized** *average simulations are shown in (a) and the comparison of the cross-sections between wells is shown in (b) for a* **5x5** cell *pattern on test problem #4.*

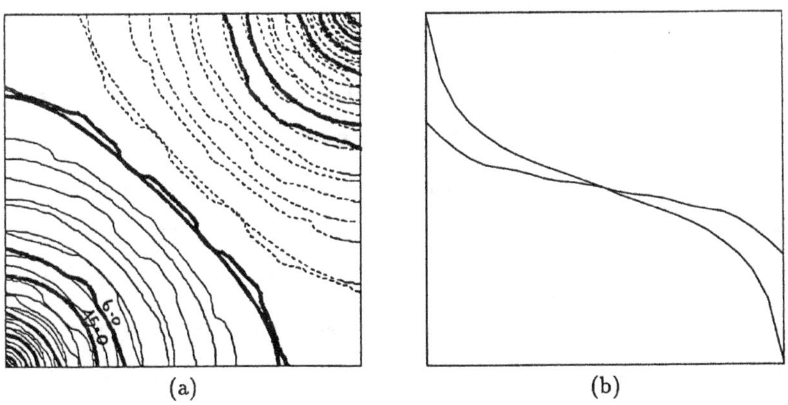

FIG. 24. *The pressure contours for the heterogeneous and* **harmonic** *average simulations are shown in (a) and the comparison of the cross-sections between wells is shown in (b) for a* **5x5** cell *pattern on test problem #4.*

(a)

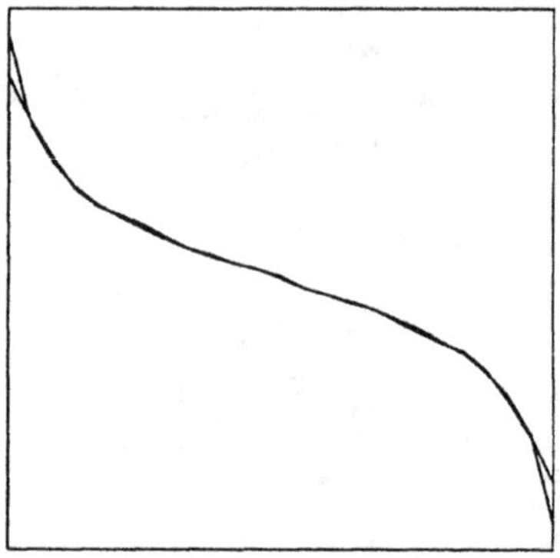

(b)

FIG. 25. *The pressure contours for the heterogeneous and geometric average simulations are shown in (a) and the comparison of the cross-sections between wells is shown in (b) for a 5x5 cell pattern on test problem #4.*

6. Conclusions. Although homogenization theory does not take into account boundary layers, all our test cases have been run on quarter five-spot patterns which have real boundary layers near the boundaries of the domain and near the wells. These boundary layers, however, have not effected the results of the simulations adversely.

In the same way one main assumption is that the size of the heterogeneities is on the same order as the cell size ϵ. But in test problem #4, the cell size is $\epsilon = \frac{1}{5}$ and the "fissures" have dimension $\frac{1}{10}$ the cell size i.e. $\frac{1}{50}$. Even though the heterogeneity is relatively small compared with ϵ one should notice that the homogenized results are in good agreement with the heterogeneous results.

Because all of the simulations have been done with a constant injection rate of $q_0 = 0.25$bbl/day, the saturation contours are not very sensitive to any change in the absolute permeability, but the pressure field does change dramatically. Thus the trus test in the results is in how well the homogenized pressure fields match the heterogeneous pressure field. In the results presented the best match of the pressure fields is obtained by using the homogenization theory presented in this paper. The other averaging methods produce inaccurate pressure fields.

Finally, taking into account previous numerical simulations reported in [2] and [3], on core samples and stratas, including various Peclet numbers and strong boundary effects one can conclude that the homogenization technique presented here is a robust method.

Acknowledgements: This work has been partially supported by a U.S.-France Cooperative Science Program (NSF/CNRS). We would like to also thank the Institute for Scientific Computation at the University of Wyoming and Chevron Oil Field Research Company for allowing us to use their computational facilities in this work.

REFERENCES

[1] M.B. Allen, Ewing R.E., and J.V. Koebbe. Mixed finite-element methods for computing groundwater velocities. In *International Conference Series on Advances in Numerical Methods in Engineering Theory and Applications, ed. Zienkiewicz, O.C., et. al.*, Swansea, U.K., January 7-11 1985.

[2] B. Amaziane. *Application des Techniques d'Homogénéisation aux Ecoulements Diphasiques Incompressibles en Milieu Poreux.* PhD thesis, University of Lyon 1, 1988.

[3] B. Amaziane and A. Bourgeat. Effective behavior of two-phase flow in heterogeneous reservoir. *IMA*, 11:1–22, 1988.

[4] B. Amaziane and T. Dumont. Calcul de coefficients homogénéisés : implémentation dans modulef et résultats numériques. *Publications du laboratoire d'Analyses Numérique no. 61 U.A. 740 Lyon, Saint-Etienne*, 1987.

[5] Y. Amirat, K. Hamdache, and A. Ziani. Homogénéisation d'équations hyperboliques du premier ordre, application aux milieux poreux. *INRIA*, 803, 1988.

[6] Y. Amirat, K. Hamdache, and A. Ziani. Homogénéisation d'un modéle d'écoulements miscibles en milieux poreux. *INRIA*, 802, 1988.

[7] A. Bensoussan, J.L. Lions, and G. Papanicolaou. *Asymptotic Analysis for Periodic Structures.* North-Holland, 1978.

[8] A. Bourgeat. Global behavior of two-phase flow simulation. In *3rd International Conference on Boundary Layers*, pages 157–160, Boole Press Pub., Dublin, 1985.

[9] A. Bourgeat. Homogenization method applied to the behavior of a naturally fissured reservoir. In K.I. Gross, editor, *Mathematical Methods in Energy Research*, pages 181–193, SIAM, 1984.

[10] A. Bourgeat. Homogenization of two-phase flow. In *the Institute on Nonlinear Analysis and Applications, Symposia in Pure Mathematics, ed. F. Browder*, AMS, 1985.

[11] A. Bourgeat. Homogenized behavior of two-phase flow in a naturally fissured reservoir with uniform fractures. *Computer Methods in Applied Mechanics and Engineering*, (47):205–217, 1984.

[12] A. Bourgeat. Nonlinear homogenization of two-phase flow in inhomogeneous media. In *Lecture Notes in Pure and Applied Mathematics, Vol. 45.*, pages 157–163, Marcel Dekker, New York, 1985.

[13] A. Bourgeat, M. Quintard, and S. Whitaker. Eléments de comparaison entre la méthode d'homogénéisation et la méthode de prise de moyenne avec fermeture. *C.R.A.S. Série II*, 360:463–466, 1988.

[14] W.T. Cardwell and R.L. Parsons. Average permeabilities of heterogeneous oil sands. *Trans,. AIME*, 160.

[15] H.C. Chang. Multiscale analysis of effective transport in periodic heterogeneous media. *Chem. Eng. Comm.*, 15:83–91, 1982.

[16] G. Chavent. A new formulation of diphasic incompressible flows in porous media. In *Lecture Notes in Mathematics Vol. 503*, pages 258–270, Springer-Verlag, 1976.

[17] G. Chavent, B. Cockburn, G. Cohen, and J. Jaffre. Une méthode d'eléments finis pour la simulation dans un réservoir de déplacements bidimensionnels d'huile par de l'eau. *INRIA*, 353, 1985.

[18] G. Chavent, G. Cohen, and J. Jaffre. Discontinuous upwinding and mixed finite elements for two-phase flows in reservoir simulation. *Computer Methods In Applied Mechanics and Engineering*, (47):93–118, 1984.

[19] G. Chavent and J. Jaffre. *Mathematical Models and Finite Elements for Reservoir Simulation.* North-Holland, Amsterdam, 1986.

[20] G. Chavent, J. Jaffre, R. Eymard, D. Guerillot, and L. Weill. Discontinuous and mixed finite elements for two-phase incompressible flow. In *1987 SPE Reservoir Simulation Symposium*, SPE, San Antonio, Texas, 1987. SPE Paper 16018.

[21] R.E. Ewing. Problems arising in the modeling of processes for hydrocarbon recovery. In R.E. Ewing, editor, *Mathematics of Reservoir Simulation, Frontiers in Applied Mathematics Vol. 1*, pages 3–34, SIAM, 1983.

[22] R.E. Ewing and M.F. Wheeler. *Computational Aspects of Mixed Finite Element Methods.* North-Holland, 1983.

[23] J.V. Koebbe. *Numerical Schemes for the Immiscible Displacement Equations Using a General Polynomial Framework for the Saturation Equation and Mixed Finite Element Methods for the Pressure.* PhD thesis, University of Wyoming, August 1988.

[24] S.N. Kruzkov and S.M. Sukorjànskii. Boundary value problems for systems of two-phase porous flow type: statement of th problems, questions of solvability, justification of approximation methods. *Math. USSR, Sbornik*, 33(1):62–80, 1977.

[25] G. Matheron. *Eléments pour une Théorie des Milieux Poreux.* Masson, Paris, 1967.

[26] F. Murat. H-convergence. 1977–78. Séminaire d'Analyse Fonctionnelle et Numérique, Université d'Alger.

[27] G. Papanicolaou and S. Varadan. Diffusion in region with many small holes. In *Lecture Notes in Control and Information, Vol. 75*, pages 190–206, Springer-Verlag, 1980.

[28] D.W. Peaceman. *Fundamentals of Numerical Reservoir Engineering.* Elsevier Scientific Publishing, 1977.

[29] E. Sanchez-Palencia. *Nonhomogeneous Media and Vibration Theory. Lecture Notes in Physics, Vol. 127.* Springer-Verlag, 1980.

[30] L. Tartar. Incompressible fluid flow in a porous medium. convergence of the homogenization process. 1980. Appendix to book by E. Sanchez-Palencia above.

[31] J.E. Warren and F.F. Skiba. Macroscopic dispersion. *SPEJ*, 215–230, 1964.

Transport in Porous Media **6**: 549–565, 1991
© 1991 *Kluwer Academic Publishers. Printed in the Netherlands.*

A LIMIT FORM OF THE EQUATIONS FOR IMMISCIBLE DISPLACEMENT IN A FRACTURED RESERVOIR

JIM DOUGLAS, JR.,[1] PAULO JORGE PAES-LEME,[2] AND JEFFREY L. HENSLEY[3]

Abstract. We study a model for simulating the flow of an immiscible displacement (waterflooding) of one incompressible fluid by another in a naturally fractured petroleum reservoir when the matrix blocks are quite small. This model is equivalent to a transformed one for immiscible flow in an unfractured reservoir with a reduced saturation and a saturation-dependent porosity. Existence and uniqueness of classical solutions are established. We present some numerical results and a comparison with a single porosity model.

Key words. Immiscible displacement, fractured reservoir, single porosity model.

1. Introduction. Naturally fractured reservoirs are formed by stress forces that occur in slow deformations of the earth's crust, with fractures frequently appearing in a reasonably regular pattern. Though the void volume in the fractures is small in comparison to that in the matrix blocks of the sedimentary formation, the very high permeability associated with the fractures requires that flow in fractured reservoirs be modelled so as to take the dual structure into account. Several models, each appropriate under different assumptions on the geometry of the fractures have been developed; see the survey article [**6**] for a description of a collection of models on two-sheeted coverings of the domain (reservoir). Here, we consider the case when the dimensions of the matrix blocks are very small. In this situation, we can neglect the viscous and gravity forces in the flow interior to the matrix blocks; therefore, we assume that the only forces that act on the blocks are the capillary forces and that these forces are in equilibrium across the surface between a block and the surrounding fractures. The model proposed in this paper is obtained as a limit when the size of the matrix blocks tends to zero in the two-sheeted models studied in [**3**], [**4**], [**6**], and [**7**]. The limiting process here is based on physical considerations; the justification for the resulting model is obtained through the convergence of the recovery curves for the medium and small block models of [**6**], [**7**], and [**4**] to that computed for the limit model of this paper. It should be noted, in particular, that the limit model is not derivable by some type of homogenization or averaging procedure beginning from the medium or small block models. The differential equations for the limiting model can be transformed to be equivalent to those for an unfractured reservoir, except that a reduced saturation must be introduced to account for the addition of a nonlinear capacitance term.

This paper is organized in the following way. In §2 we introduce the limit model and determine some of its properties. We prove existence and uniqueness of classical solutions in §3; uniqueness is proved in $\mathbf{R}^3$ and existence in $\mathbf{R}^2$. (Existence in $\mathbf{R}^3$ for a weak solution has quite recently been proved by Arbogast [**2**]; thus, we have made no effort to find a second proof for the existence of a weak solution in this case.) Then, §4 is devoted to the presentation of some numerical simulations; these calculations are intended to provide the

[1] Department of Mathematics, Purdue University, West Lafayette, IN 47907.
[2] Instituto Politécnico do Rio de Janeiro, 28600 Nova Friburgo, R.J., and Departamento de Matemática, Pontifícia Universidade Católica do R.J., 22453 Rio de Janeiro, R.J., Brasil.
[3] Department of Mathematics, Purdue University, West Lafayette, IN 47907.

justification of the model mentioned above and to illustrate some parametric dependencies of the model. An abbreviated derivation of the global pressure formulation of the equations describing incompressible, immiscible flow in an ordinary reservoir is given in §5. Finally, a nomenclature is given in §6.

2. The Model. We begin with the medium block model of [6], [7], and [8] that is based on three families of parallel, equispaced, planar fractures. This model, which can be derived from a microscopic description of ordinary two-phase, incompressible, immiscible flow in a porous medium, is formulated on a two-sheeted covering of the physical reservoir, with the fracture sheet Ω coinciding with the reservoir. Over each point $x \in \Omega$ a matrix block Ω_x is attached to the fracture sheet; these blocks are topologically distinct from each other. In general, these blocks have a rhomboidal shape.

Following [6] and [8], the flow in the fracture system is governed by the following set of partial differential equations:

$$\Phi\frac{\partial S_w}{\partial t} - \nabla_x \cdot [\frac{KK_{rw}}{\mu_w}(\nabla_x P_w + \rho_w g \nabla_x z)] = Q_{w,e} + Q_{w,m}, \quad x \in \Omega, t > 0, \tag{2.1}$$

$$\Phi\frac{\partial S_o}{\partial t} - \nabla_x \cdot [\frac{KK_{ro}}{\mu_o}(\nabla_x P_o + \rho_o g \nabla_x z)] = Q_{o,e} + Q_{o,m}, \quad x \in \Omega, t > 0, \tag{2.2}$$

$$S_w + S_o = 1,$$

$$P_o - P_w = P_c(S_o),$$

where $Q_{w,e}$ and $Q_{o,e}$ are the volumetric external flow rates (at wells in practice) of the water and oil phases, respectively. The terms $Q_{w,m}$ and $Q_{o,m}$ describe the volumetric flow rates of the water and oil phases resulting from flow out of the matrix blocks into the fractures. As in [4], we prefer to use a global pressure formulation [1], [9], [5] of the equations, which we derive below.

Let

$$S = S_o = 1 - S_w. \tag{2.3}$$

Define the total mobility to be

$$\Lambda(S) = \frac{K_{ro}(S)}{\mu_o} + \frac{K_{rw}(S)}{\mu_w} \tag{2.4}$$

and the phase mobilities to be

$$\Lambda_\theta(S) = \frac{K_{r\theta}(S)}{\mu_\theta \Lambda(S)}, \quad \theta = o, w. \tag{2.5}$$

Define the global pressure to be

$$P = \frac{1}{2}\{P_o + P_w + \int_0^{P_c} (\Lambda_o - \Lambda_w)(P_c^{-1}(\xi))d\xi\}, \tag{2.6}$$

the global density to be

$$\rho_f(S) = \Lambda_o(S)\rho_o + \Lambda_w(S)\rho_w, \tag{2.7}$$

and the total fluid velocity to be

$$U = -K\Lambda(\nabla_x P + \rho_f g \nabla_x z). \tag{2.8}$$

It is shown in §5 (and in [**5**], [**9**]) that (2.1)-(2.8) lead to the global pressure equation

$$\nabla_x \cdot U = Q_e = Q_{w,e} + Q_{o,e} \tag{2.9}$$

and, assuming that $Q_{o,e} = -\Lambda_o Q_e^+ = -\Lambda_o max(0, Q_e)$, a saturation equation

$$\Phi\frac{\partial S}{\partial t} + \Lambda_o' U \cdot \nabla_x S - \nabla_x \cdot (K\Lambda\Lambda_o\Lambda_w P_c' \nabla_x S) - \nabla_x \cdot (K\Lambda\Lambda_o\Lambda_w(\rho_o - \rho_w)g\nabla_x z) = -\Lambda_o Q_e^+ + Q_{o,m}. \tag{2.10}$$

The initial saturation $S(x,0)$ and boundary conditions must be specified. For simplicity, assume that no flow of either phase occurs across the external boundary $\partial\Omega$. It follows that

$$U \cdot \nu_{\partial\Omega} = 0. \tag{2.11}$$

An additional boundary condition will be specified below in (2.33).

Denote the coordinates on the matrix block Ω_x by (x, y, t), where y is the local spatial coordinate on the block. The flow on Ω_x is described by the following differential equations:

$$u = -k\lambda(\nabla_y p + \rho_m g \nabla_y z), \quad y \in \Omega_x, t > 0, \tag{2.12}$$

$$\nabla_y \cdot u = 0, \quad y \in \Omega_x, t > 0, \tag{2.13}$$

and

$$\phi\frac{\partial s}{\partial t} + \lambda_o' u \cdot \nabla_y s - \nabla_y \cdot (k\lambda\lambda_o\lambda_w p_c' \nabla_y s) - \nabla_y \cdot (k\lambda\lambda_o\lambda_w(\rho_o - \rho_w)g\nabla_y z) = 0, \tag{2.14}$$

where

$$s = s_o = 1 - s_w, \tag{2.15}$$

$$\lambda(s) = \frac{k_{ro}(s)}{\mu_o} + \frac{k_{rw}(s)}{\mu_w}, \tag{2.16}$$

$$\lambda_\theta(s) = \frac{k_{r\theta}(s)}{\mu_\theta \lambda(s)}, \quad \theta = o, w, \tag{2.17}$$

$$p_o - p_w = p_c(s_o), \tag{2.18}$$

$$p = \frac{1}{2}\{p_o + p_w + \int_0^{p_c} (\lambda_o - \lambda_w)(p_c^{-1}(\xi))d\xi\}, \tag{2.19}$$

$$\rho_m(s) = \lambda_o(s)\rho_o + \lambda_w(s)\rho_w. \tag{2.20}$$

The initial saturation in each matrix block is assumed to be consistent with capillary equilibrium between the fractures and the block; that is,

$$s(x,y,0) = (p_c^{-1} \circ P_c)(S(x,0)), \quad x \in \Omega, y \in \Omega_x. \tag{2.21}$$

At the boundary of the blocks, the oil saturations satisfy the relation

$$s(x,y,t) = p_c^{-1}(P_c(S(x,t))) = (p_c^{-1} \circ P_c)(S(x,t)), \quad x \in \Omega, y \in \partial\Omega_x, t > 0, \tag{2.22}$$

and

$$p(x,y,t) = P(x,t), \quad x \in \Omega, y \in \partial\Omega_x, t \geq 0. \tag{2.23}$$

It has been shown in [**8**] that gravity can be neglected in the blocks for small blocks. The small block model described in [**6**] results from ignoring exactly these gravitational effects, with a consequent simplification in the differential system on the blocks. When the gravity term in (2.12) is omitted, (2.13) reduces to a homogeneous elliptic equation for p on Ω_x; since the boundary values given by (2.23) for p are constant for any fixed t,

$$p(x,y,t) = P(x,t), \qquad y \in \Omega_x. \tag{2.24}$$

Then, it follows that

$$u(x,y,t) = 0, \qquad y \in \Omega_x. \tag{2.25}$$

Thus, pressure differences are neglected inside each matrix block, and the only relevant mechanism on a block is the effect of the capillary forces. The equations (2.12)-(2.14) governing the flow in the matrix blocks are then reduced to the single equation

$$\phi\frac{\partial s}{\partial t} - \nabla_y \cdot (k\lambda\lambda_o\lambda_w p_c' \nabla_y s) = 0, \tag{2.26}$$

subject to the boundary condition (2.22). The fracture equations (2.8)-(2.10) are coupled to (2.26) via the term

$$Q_{o,m}(x,t) = -\frac{1}{|\Omega_x|}\int_{\partial\Omega_x} k\lambda\lambda_o \frac{\partial p_o}{\partial \nu} d\eta, \quad x \in \Omega, \quad t > 0. \tag{2.27}$$

Using Green's identity and conservation of mass in the oil phase, equation (2.27) can be rewritten as

$$Q_{o,m}(x,t) = -\frac{1}{|\Omega_x|}\int_{\Omega_x} \phi\frac{\partial s}{\partial t} dy, \quad x \in \Omega_x, \quad t > 0. \tag{2.28}$$

This model should be applicable for a small, but not necessarily vanishingly small, block.

The object now is to find a limit form for the model as the blocks tend to zero size; that is, as $diam(\Omega_x) \to 0$, uniformly for all $x \in \Omega$. We assume that the void space remains distributed in a fixed way; i.e., the porosities ϕ of the matrix blocks and Φ of the fractures are independent of the limit process. Similarly, assume that the relative and total permeabilities in the fractures remain fixed, as well as the capillary pressure relations in both the fractures and the matrix blocks. The relaxation time for the saturation equation (2.12) in each of these small blocks will be much smaller than that for the fractures; this is a statement of physical fact, though it can be verified in a semimathematical fashion by computing the first eigenvalue in a separation of variables solution for the differential systems that would result from setting the coefficients in (2.8)-(2.10) and (2.12)-(2.14) to their average values and noticing the very large difference in the expected extents of the fracture domain and a typical matrix block. As a consequence, we can assume that the delay in response in the saturation of a block can be ignored, and the solution of (2.25), (2.21) can be approximated very well by the relation

$$s(x,y,t) = (p_c^{-1} \circ P_c)(S(x,t)), \quad y \in \Omega_x, x \in \Omega; \tag{2.29}$$

note that this is the statement of the physical reality that the time scale in the blocks is sufficiently faster than that in the fractures that equilibrium is reached in a block in a time increment that is less than an increment of physical interest in the fractures. Under this hypothesis, it follows that

$$Q_{o,m}(x,t) \to -\phi \frac{\partial}{\partial t}[p_c^{-1} \circ P_c(S(x,t))], \quad x \in \Omega. \tag{2.30}$$

Substitution of the limit for $Q_{o,m}(x,t)$ into (2.10) gives the modified saturation equation

$$\begin{aligned} &\frac{\partial}{\partial t}[\Phi S + \phi p_c^{-1} \circ P_c(S)] + \Lambda_o' U \cdot \nabla_x S - \nabla_x \cdot (K\Lambda\Lambda_o\Lambda_w P_c' \nabla_x S) \\ &\quad - \nabla_x \cdot (K\Lambda\Lambda_o\Lambda_w(\rho_o - \rho_w) g \nabla_x z) = -\Lambda_o Q_e^+. \end{aligned} \tag{2.31}$$

The limit model is given by the pressure and saturation equations (i.e., by the system (2.8), (2.9) and (2.31)), subject to an initial condition

$$S(x,0) = S^0(x) \tag{2.32}$$

and the (typical) boundary conditions (2.11) and

$$K\Lambda\Lambda_o\Lambda_w[P_c'\nabla_x S + (\rho_o - \rho_w)g\nabla_x z] \cdot \nu_{\partial\Omega} = 0. \tag{2.33}$$

Equation (2.33) results from differencing weighted multiples of the Darcy velocities for oil and water and requiring the flow across $\partial\Omega$ to vanish for this combination.

While this model was derived from models that impose restrictions of local periodicity on the families of fractures defining the blocks, the limit itself can be given a very reasonable interpretation under the sole constraint that the blocks, of whatever shape, have small diameter. The absolute permeability tensor in the fractures can be derived from the micropermeability properties of the fractures and the blocks and the geometry of the fracture families in the periodic case (see **[3]**, **[6]**); it would have to be determined entirely experimentally in the general case.

3. Existence, Uniqueness and Regularity. In this section we assume that there exists a small positive constant ϵ such that

$$\begin{aligned} &min\, K_{ro}, \quad min\, K_{rw} \geq \epsilon > 0, \\ &max\, |P_c'| \leq 1/\epsilon. \end{aligned}$$

This is not an inherent property of the model, but it will permit us to look at a nondegenerate mathematical problem. These conditions hold if the saturation remains in an interval bounded away from the residual saturations for both phases.

Set

$$\varphi(S) = \Phi S + \phi(p_c^{-1} \circ P_c)(S), \quad 0 \leq S \leq 1. \tag{3.1}$$

Rewrite the pressure equation (2.8)–(2.9) in the form

$$-\nabla_x \cdot (\gamma \nabla_x P + \delta \nabla_x z) = Q_e, \tag{3.2}$$

where

$$\begin{aligned} \gamma &= \gamma(x, S) = K\Lambda, \\ \delta &= \delta(x, S) = K\Lambda\rho_f g. \end{aligned}$$

The boundary condition becomes

$$(\gamma \nabla_x P + \delta \nabla_x z) \cdot \nu_\Omega = 0, \quad x \in \partial\Omega. \tag{3.3}$$

The saturation equation (2.31) becomes

$$\frac{\partial}{\partial t}\varphi(S) - \nabla_x \cdot [\alpha \nabla_x S + \beta U + \zeta] = Q_{o,e}, \tag{3.4}$$

where

$$\begin{aligned} \alpha &= \alpha(x, S) = K\Lambda\Lambda_o\Lambda_w P_c', \\ \beta &= \quad \beta(S) = -\Lambda_o(S), \\ \zeta &= \zeta(x, S) = K\Lambda\Lambda_o\Lambda_w(\rho_o - \rho_w) g \nabla_x z; \end{aligned}$$

the boundary condition becomes

$$(\alpha \nabla_x S + \zeta) \cdot \nu = 0, \quad x \in \partial\Omega. \tag{3.5}$$

Now, consider semi-classical solutions in the sense defined by Kružkov and Sukorjanskiĭ [9]; $(\cdot,\cdot)$ denotes the inner product in $L^2(\Omega)$. This definition is applicable for either two-dimensional or three-dimensional domains Ω and is equivalent to the one given in [9].

DEFINITION. *A pair of functions* $\{P(x,t), S(x,t)\}$, *bounded and measurable in* $\overline{\Omega \times J}$, $J = [0, T]$, *is called a semi-classical solution of the problem (3.2)–(3.5) if*

i) P *and* S *are Lipschitz-continuous in* $\Omega \times J$;

ii) $(\gamma \nabla_x P + \delta \nabla_x z, \nabla_x q) = (Q_e, q),\ q \in H^1(\Omega), t \in J;$

iii) $(\frac{\partial \varphi(S)}{\partial t}, v) + (\alpha \nabla_x S + \beta U + \zeta, \nabla_x v) = (Q_{o,e}(S), v),\ v \in H^1(\Omega),\ t \in J.$

Note that $Q_e = Q_e(x,t)$ is given and it does not depend on the solution, whereas $Q_{o,e}$ does.

Let us consider uniqueness; the theorem below is valid for Ω a bounded domain in either $\mathbf{R}^2$ or $\mathbf{R}^3$.

THEOREM 1. *There exists at most one semi-classical solution of the problem (3.2)–(3.5) satisfying a given initial condition (2.32).*

PROOF: Let $\{P_1, S_1\}$ and $\{P_2, S_2\}$ be solutions for the same data $S^0(x)$ and $Q_e(x,t)$. Set

$$\Pi = P_1 - P_2 \quad \text{and} \quad \Sigma = S_1 - S_2.$$

Then, from *ii)* in the definition above,

$$(\gamma_1 \nabla_x \Pi, \nabla_x q) + ((\delta_1 - \delta_2)\nabla_x z, \nabla_x q) + ((\gamma_1 - \gamma_2)\nabla_x P_2, \nabla_x q) = 0,\ q \in H^1(\Omega).$$

Note that, since Λ is a continuously differentiable function of S, there exists $b \in L^\infty(\Omega \times J)$ such that

$$(\gamma_1 - \gamma_2)\nabla_x P_2 + (\delta_1 - \delta_2)\nabla_x z = -b(x,t)\Sigma.$$

Thus,

$$(\gamma_1 \nabla_x \Pi, \nabla_x q) = (b\Sigma, \nabla_x q), \quad q \in H^1. \tag{3.6}$$

A useful equation for the difference in saturations will require more effort. First,

$$\begin{aligned}(\frac{\partial \varphi(S_1)}{\partial t} - \frac{\partial \varphi(S_2)}{\partial t}, v) + (\alpha_1 \nabla_x \Sigma + \beta_1(U_1 - U_2) + (\zeta_1 - \zeta_2)\\ + (\alpha_1 - \alpha_2)\nabla_x S_2 + (\beta_1 - \beta_2)U_2, \nabla_x v) = (Q_{o,e}(S_1) - Q_{o,e}(S_2), v)\end{aligned}$$

Note that there exist functions c, d, and e in $L^\infty(\Omega \times J)$ such that

$$\begin{aligned}(\alpha_1 - \alpha_2)\nabla_x S_2 + \beta_1(U_1 - U_2) + (\beta_1 - \beta_2)U_2 + (\zeta_1 - \zeta_2) = -c(x,t)\Sigma - d(x,t)\nabla_x \Pi,\\ Q_{o,e}(S_1) - Q_{o,e}(S_2) = e(x,t)\Sigma.\end{aligned}$$

Thus,

$$(\frac{\partial \varphi(S_1)}{\partial t} - \frac{\partial \varphi(S_2)}{\partial z}, v) + (\alpha_1 \nabla_x \Sigma, \nabla_x v) = (c\Sigma + d\nabla_x \Pi, \nabla_x v) + (e\Sigma, v),\ v \in H^1. \tag{3.7}$$

Now, we wish to choose $q = \Pi$ and $v = \Sigma$. The first term in (3.7) can be handled by use of a transformation due to Dupont and Wheeler [10], §3.4. Let $R : \mathbf{R}^2 \to \mathbf{R}$ be defined by

$$R(S,\xi) = \int_0^\xi \frac{\partial\varphi}{\partial S}(S-\mu)\mu d\mu.$$

Note first that

$$\int_\Omega R(S,\xi)dx = \int_\Omega \int_0^\xi \frac{\partial\varphi}{\partial S}(S-\mu)\mu d\mu\, dx \geq \int_\Omega \varphi'_{\min}\frac{1}{2}\xi^2 dx \geq \frac{1}{2}\varphi'_{\min}\|\xi\|_0^2.$$

Next, consider $\frac{d}{dt}\int_\Omega R(S_1, S_1 - S_2)dx$:

$$\begin{aligned}\frac{d}{dt}\int_\Omega R(S_1,S_1-S_2)dx &= \int_\Omega \varphi'(S_2)\frac{\partial\Sigma}{\partial t}\Sigma dx + \int_\Omega\int_0^\Sigma \varphi''(S_1-\mu)\frac{\partial S_1}{\partial t}\mu d\mu dx\\ &= \int_\Omega \left[\varphi'(S_1)\frac{\partial S_1}{\partial t} - \varphi'(S_2)\frac{\partial S_2}{\partial t}\right]\Sigma dx\\ &\quad + \int_\Omega [\varphi'(S_2)-\varphi'(S_1)]\frac{\partial S_1}{\partial t}\Sigma dx + \int_\Omega\int_0^\Sigma \varphi''(S_1-\mu)\frac{\partial S_1}{\partial t}\mu d\mu dx.\end{aligned}$$

Hence,

$$\begin{aligned}\int_\Omega \left[\frac{\partial\varphi(S_1)}{\partial t} - \frac{\partial\varphi(S_2)}{\partial t}\right]\Sigma dx = \frac{d}{dt}\int_\Omega R(S_1,\Sigma)dx + \int_\Omega [\varphi'(S_1)-\varphi'(S_2)]\frac{\partial S_1}{\partial t}\Sigma dx\\ - \int_\Omega\int_0^\Sigma \varphi''(S_1-\mu)\frac{\partial S_1}{\partial t}\mu d\mu dx.\end{aligned}$$

Thus,

$$\begin{aligned}&\frac{d}{dt}\int_\Omega R(S_1,\Sigma)dx + (\alpha_1\nabla_x\Sigma,\nabla_x\Sigma)\\ &\quad = (c\Sigma + d\nabla_x\Pi,\nabla_x\Sigma) + (e\Sigma,\Sigma) + \int_\Omega\int_0^\Sigma \varphi''(S_1-\mu)\frac{\partial S_1}{\partial t}\mu d\mu dx\\ &\qquad - \int_\Omega [\varphi'(S_1)-\varphi'(S_2)]\frac{\partial S_1}{\partial t}\Sigma dx\\ &\quad \leq M_1\|\Sigma\|_0^2 + M_2\|\nabla_x\Pi\|_0^2 + \frac{1}{2}\alpha_{min}\|\nabla_x\Sigma\|_0^2.\end{aligned}$$

Since it follows from (3.6) that

$$\|\nabla_x\Pi\|_0 \leq M_3\|\Sigma\|,$$

then

$$\frac{d}{dt}\int_\Omega R(S_1,\Sigma)dx + \frac{1}{2}(\alpha_1\nabla_x\Sigma,\nabla_x\Sigma) \leq M_4\|\Sigma\|_0^2.$$

The inequality $\int_\Omega R(S_1,\Sigma)dx \geq \frac{1}{2}\varphi'_{min}\|\Sigma\|_0^2$ and a Gronwall argument show that

$$\|\Sigma(t)\|_0 + \|\Sigma\|_{L^2(0,t;H^1)} + \|\Pi\|_{L^2(0,t;H^1)} \leq M(t)\|\Sigma(0)\|_0 = 0,$$

and uniqueness of the semi-classical solution has been established.

A trivial modification in the argument above gives continuous dependence on the data for semi-classical solutions. With Σ and Π the differences defined above, it follows that

$$\begin{aligned}\|\Sigma\|_{L^\infty(J;L^2(\Omega))} &+ \|\Sigma\|_{L^2(J;H^1(\Omega))} + \|\Pi\|_{L^2(J;H^1(\Omega))} \\ &\leq M(T)\Big[\|S_1^0 - S_2^0\|_0 + \|Q_{1,e} - Q_{2,e}\|_{L^2(J;L^2(\Omega))}\Big].\end{aligned} \tag{3.8}$$

The existence and regularity of the solution of the system given by (2.8), (2.9), and (2.31) will be demonstrated only in the case of a *two-dimensional* domain; see **[2]** for a proof of the existence of a particular type of weak solution in the three-dimensional case. Here, let us assume that the function φ of (3.1) is smooth (i.e., $\varphi \in C^2([0,1])$, at least). Since $\varphi' \geq \Phi$, φ is invertible. Let

$$C = \varphi(S), \tag{3.9}$$

so that (3.4) can be rewritten as

$$\frac{\partial C}{\partial t} - \nabla_x \cdot (\tilde{\alpha}\nabla_x C + \tilde{\beta}U + \tilde{\zeta}) = -\tilde{\Lambda}_0 Q_e^+, \tag{3.10}$$

where

$$\begin{aligned}\tilde{\alpha} &= \tilde{\alpha}(x,C) = \alpha(x,S)/\varphi'(S),\\ \tilde{\beta} &= \tilde{\beta}(C) = \beta(S),\\ \tilde{\zeta} &= \tilde{\zeta}(x,C) = \zeta(x,S),\\ \tilde{\Lambda}_i &= \tilde{\Lambda}_i(x,C) = \Lambda_i(x,S);\end{aligned}$$

if either of the porosities is spatially dependent, then the functions $\tilde{\alpha}$ and $\tilde{\zeta}$ will have slightly more complicated dependencies on x than indicated above, but for porosities bounded away from zero the form of (3.10) will remain the same.

Equations (2.8) and (2.9) can be written analogously in terms of C in place of S:

$$U = -\tilde{\gamma}\nabla_x P - \tilde{\delta}\nabla_x z, \tag{3.11}$$

$$\nabla_x \cdot U = Q_e, \tag{3.12}$$

where

$$\begin{aligned}\tilde{\gamma} &= \tilde{\gamma}(x,C) = \gamma(x,S),\\ \tilde{\delta} &= \tilde{\delta}(x,C) = \delta(x,S).\end{aligned}$$

The resulting system is of exactly the same form as that for a standard, unfractured immiscible displacement, except that C belongs to a different, still compact, interval. If initial values

$$C(x,0) = C^0(x), \ x \in \Omega, \tag{3.13}$$

and the boundary conditions

$$U \cdot \nu = -(\tilde{\alpha}\nabla_x C + \tilde{\zeta}) \cdot \nu = 0, \ x \in \partial\Omega, \tag{3.14}$$

are imposed, then the existence and regularity results of Kružkov and Sukorjanskiĭ [**9**] can be applied directly to show that (3.10)-(3.14) has a unique classical solution for smooth, consistent data.

Finally, invert C to find a function S that is a classical solution of the original problem. Since it is clearly a semi–classical solution as well, it is the unique solution.

4. Some Numerical Experiments. While a recovery curve, the plot of the volume of the oil produced versus the volume of the water injected, gives only a gross indication of the flow of fluids within a reservoir, it is sufficient to illustrate important features of flows in naturally fractured reservoirs. Consider a rectangular reservoir Ω with height 10 meters and length 300 meters. For computational simplicity, assume that the reservoir is uniform in the other direction, so that the fracture simulation can be considered to be two-dimensional, though the matrix simulation remains three-dimensional for each block Ω_x.

Let the capillary pressure functions be assumed in the forms

$$P_c(S) = (1-S)\{\gamma(S^{-1}-1)+\theta\}, \tag{4.1a}$$

$$\begin{aligned} p_c(s) &= \alpha(\{s_0 - s + \beta\}^{\frac{1}{2}} - \beta^{\frac{1}{2}})(s - s_{rw})^{-\delta} \\ s_0 &= 1 - s_{ro}, \quad \beta = s_{ro}^2(s_0 - s_{rw})^{-2}, \end{aligned} \tag{4.1b}$$

and let the relative permeabilities be taken in the forms

$$K_{ro}(S) = 1 - S, \qquad K_{rw}(S) = S, \tag{4.2a}$$

$$k_{ro}(s) = s_0^{-2}(s_0 - s)^2, \quad k_{rw}(s) = (s - s_{rw})^2(1 - s_{rw})^{-2}. \tag{4.2b}$$

Let the base case be described by the following data:

Fluid Properties:		
Viscosity	$\mu_w = .5\ cP$	$\mu_0 = 2\ cP$
Density	$\rho_w = 1\ g/cm^3$	$\rho_0 = .7\ g/cm^3$
Absolute Permeabilities:	$K = 1\ darcy$	$k = 0.05\ darcy$
Porosities:	$\Phi = .01$	$\phi = .2$
Residual Saturations (matrix):	$s_{ro} = .15$	$s_{rw} = .2$
Capillary Pressure Parameters:		
	$\gamma = 20{,}000\ dynes/cm^2$	
	$\theta = 100\ dynes/cm^2$	
	$\alpha = 1{,}500\ dynes/cm^2$	

The residual saturations in the fractures are zero. Assume that the reservoir contains 76% oil and 24% water initially, in capillary equilibrium. For the base case, assume that the angle of inclination ϑ of the reservoir is zero. The water injection rate is $.2pv/yr$, with uniform injection along the left edge and production at the upper right corner. The numerical experiments reported below used a finite difference method that has been described in [**6**]. Tests run earlier [**6**], [**7**] showed that a 40x10 grid suffices to give accurate recovery curves. A variable time step was used, beginning with one day and ending with twenty days.

Other tests reported in [**8**] indicated quite clearly that the solution of both the medium and small block models produced recovery curves that converged to the corresponding recovery curve associated with the limit model of this paper. It was seen in [**8**] that, as the block size tends to zero, the recovery curves for the medium block model are very close to those of the limit model for block sizes below 50 cm and become almost indistinguishable from that for the limit model at about 10 cm. The recovery curves for the small block model fall in between the corresponding curves for the medium block model and the curve for the limit model. These convergences are the validations of the limit model, for the object of the model is to produce a mathematical model of greater computational simplicity than either the medium or small block models while retaining adequate engineering accuracy for very small block size. The conclusions of some of the experimental results in [**8**] differ from those reported earlier in [**6**] and [**7**]; there was a data representation error in the earlier calculations.

Figure 1 indicates that a higher absolute permeability in the fractures leads to earlier and greater water production, a conclusion that appears to be correct physically, as it is easier for water to bypass the matrix blocks. Similarly, increasing the capillary pressure function in the fractures slows oil production, as it should; see Figure 2. Correspondingly,

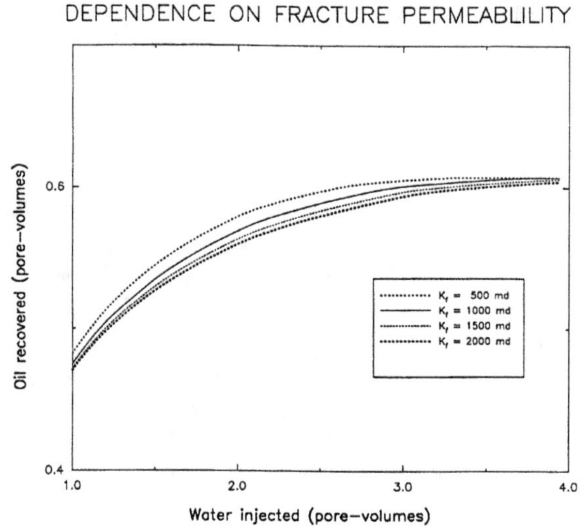

Figure 1

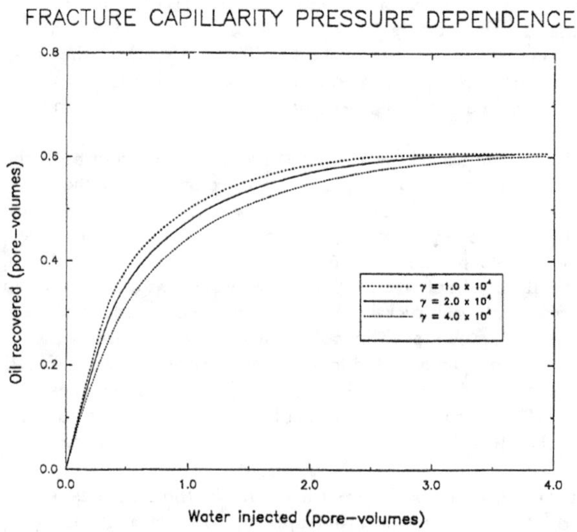

Figure 2

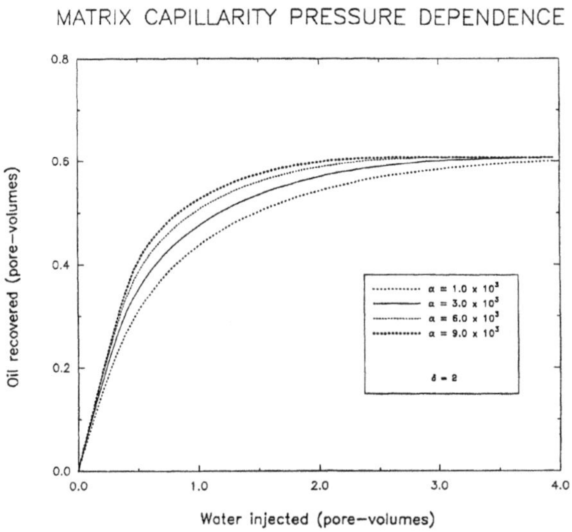

Figure 3

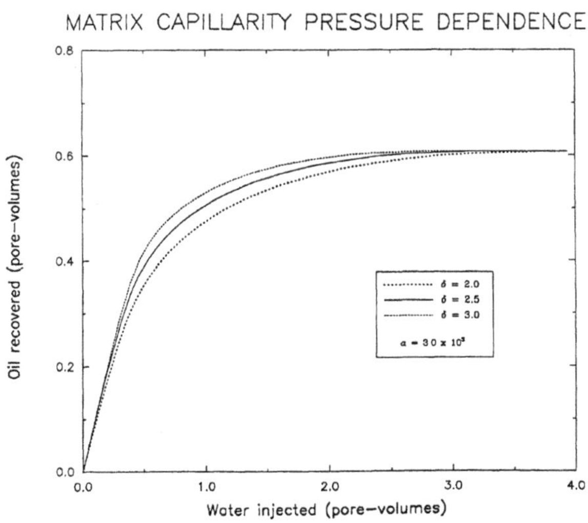

Figure 4

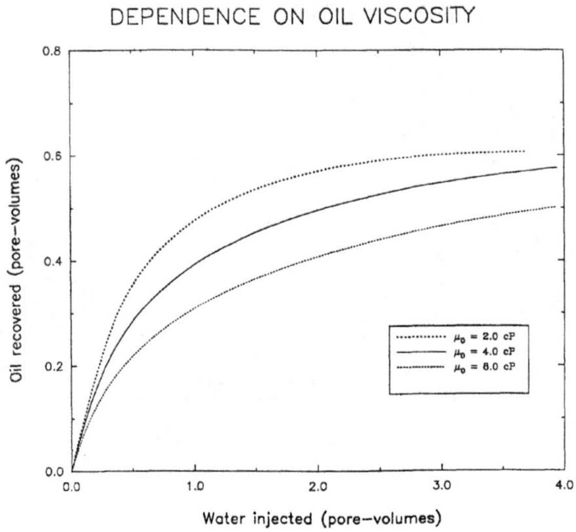

Figure 5

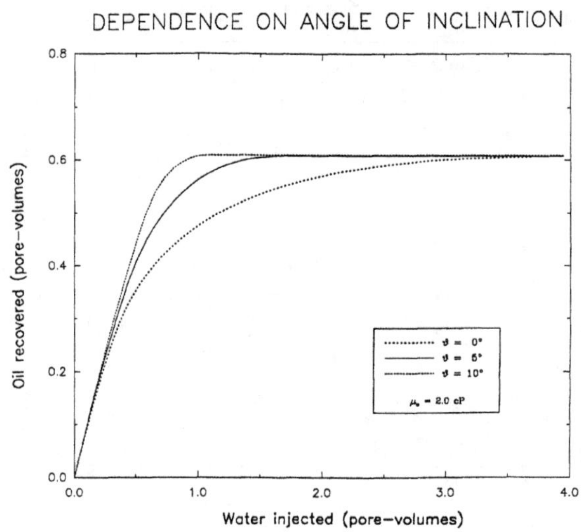

Figure 6

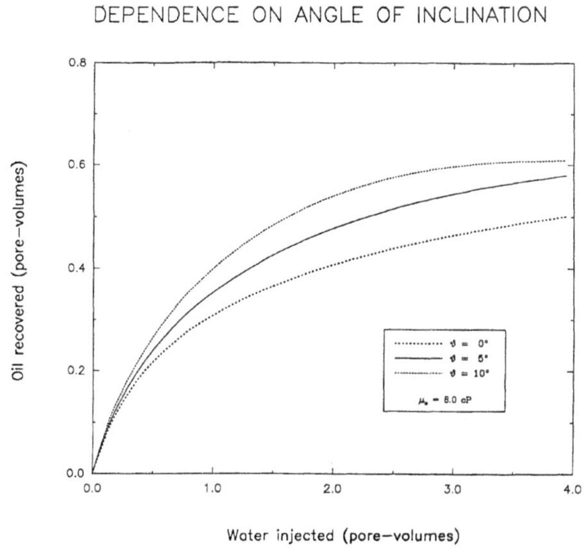

Figure 7

increasing the capillary pressure function in the matrix blocks slows the production of water and improves oil recovery; see Figures 3 and 4.

Figure 5 shows the physically reasonable behavior that oil recovery is slowed by increasing oil viscosity. If the reservoir is inclined by an angle ϑ ($\vartheta > 0$ indicating that the production corner is above the injection side of the reservoir), oil recovery is improved, the more so for more viscous oil. See Figures 6 and 7.

Additional experiments indicated a greater sensitivity to the matrix porosity when the fracture porosity was .001 than when it was .01, with more rapid oil recovery (again for *pv*–injected vs. *pv*–recovered) for a higher matrix porosity.

5. Global Pressure Formulation. The derivation of the global pressure formulation of (2.1) and (2.2) is outlined here. See also [**1**], [**9**], [**5**]. It follows easily from (2.6) that

$$\nabla_x P = \Lambda_o \nabla_x P_o + \Lambda_w \nabla_x P_w.$$

With U being given by (2.8), the pressure equation (2.9) is immediate upon addition of (2.1) and (2.2).

In order to obtain (2.10), first note that

$$\nabla_x P_o = \nabla_x P + \Lambda_w \nabla_x P_c,$$

as $\Lambda_o + \Lambda_w = 1$. Then,

$$K\Lambda\Lambda_o\nabla_x P_o = -\Lambda_o(U + K\Lambda g\rho_f\nabla_x z) + K\Lambda\Lambda_o\Lambda_w\nabla_x P_c;$$

(2.10) follows by substitution of this relation into (2.2) and using the assumption that

$$Q_{o,e} = -\Lambda_o Q_e^+.$$

6. Nomenclature.

θ - phase = o (oil) or w (water)

μ_θ - viscosity of phase θ

Φ - fracture porosity

K - absolute permeability in the fractures

$K_{r\theta}$ - relative permeability of phase θ in the fractures

P_θ - pressure of phase θ in the fractures

P_c - capillary pressure in the fractures ($= P_o - P_w$)

S_θ - saturation of phase θ in the fractures

ϕ - matrix porosity

k - absolute permeability in the matrix

$k_{r\theta}$ - relative permeability of phase θ in the matrix

p_θ - pressure of phase θ in the matrix

p_c - capillary pressure in the matrix ($= p_o - p_w$)

s_θ - saturation of phase θ in the matrix

Acknowledgements. The authors wish to thank Todd Arbogast for many discussions on the subject matter of the paper; the code used for the experiments was written in large part by him with smaller parts by the authors. We also wish to thank N. P. Nunes and Decio S. Lima, Jr., for help with the computations. The research of Douglas was supported in part by the NSF, that of Hensley in part by the U.S. Army, and that of Paes Leme in part by CNPq.

REFERENCES

1. S. N. Antoncev, *On the solvability of boundary value problems for degenerate two-phase porous flow equations*, (Russian), Dinamika Splošnoǐ Sredy Vyp. **10** (1972), 28–53.
2. Todd Arbogast, *Two–phase incompressible flow in a porous medium with various nonhomogeneous boundary conditions*, (to appear).
3. Todd Arbogast, Jim Douglas, Jr., and Ulrich Hornung, *Modeling of naturally fractured reservoirs by formal homogenization techniques*, (to appear).
4. Todd Arbogast, Jim Douglas, Jr., and Juan E. Santos, *Two-phase immiscible flow in naturally fractured reservoirs*, in "Numerical Simulation in Oil Recovery," Mary F. Wheeler, (ed.), The IMA Volumes in Mathematics and its Applications, Springer-Verlag, Berlin and New York, 1988, pp. 47–66.
5. Guy Chavent, *A new formulation of diphasic incompressible flows in porous media*, in "Applications of Methods of Functional Analysis to Problems in Mechanics," Lecture Notes in Mathematics 503, P. Germain and B. Nayroles, eds., Springer-Verlag, Berlin and New York, 1976, pp. 258–270.
6. Jim Douglas, Jr. and Todd Arbogast, *Dual-porosity models for flow in naturally fractured reservoirs*, in "Dynamics of Fluids in Hierarchial Porous Formations," J. H. Cushman, ed., Academic Press, to appear.
7. Jim Douglas, Jr., Todd Arbogast, and Paulo Jorge Paes Leme, *Two models for the waterflooding of naturally fractured reservoirs*, Paper SPE 18425, in "Proceedings, Tenth SPE Symposium on Reservoir Simulation," Society of Petroleum Engineers, Dallas, Texas, 1989, pp. 219–225.
8. Jim Douglas, Jr., Todd Arbogast, and Jeffrey L. Hensley, *A dual–porosity model for waterflooding in naturally fractured reservoirs*, to appear.
9. S. N. Kružkov and S. M. Sukorjanskiĭ, *Boundary problems for systems of equations of two-phase porous flow type; statement of the problems, questions of solvability, justification of approximate methods*, Math. USSR Sbornik **33** (1977), 62–80.
10. Mary F. Wheeler, *A priori L_2 error estimates for Galerkin approximations to parabolic partial differential equations*, SIAM J. Numer. Anal. **10** (1973), 723–759.

Keywords. porous medium, double porosity, immiscible flow, fractured reservoir.

Transport in Porous Media **6**: 567–580, 1991
© 1991 *Kluwer Academic Publishers. Printed in the Netherlands.*

Diffusion Models with Microstructure

R.E. SHOWALTER

Abstract. Models of diffusion are presented which recognize the local geometry of individual cells or storage sites and the exchange of flux on the micro-scale of these cells. Such models have been obtained by homogenization, but here we indicate stronger existence-uniqueness results of "parabolic" type can be obtained directly. Connections between these models and their historical development will be described.

Key words. Diffusion models, micro-structure of medium.

1. Distributed Microstructure Models.

We shall begin with a review of some classical models of diffusion through composite media, especially the case of a fissured medium, in which the double-porosity concept is useful. Then we introduce two models which describe also the micro-structure of the medium, the geometry of the individual cells or pores in the medium and the flux across the intricate interface which separates them from the global medium or matrix structure in which they are imbedded.

A fissured medium consists of a matrix of porous and permeable blocks or cells which are separated from each other by a highly developed system of fissures or bulk flow paths through which the majority of diffusion occurs. It is clearly impossible to determine the exact shape of the intricate interface between this cell matrix and the surrounding fissure system, and it would be equally unrealistic to attempt to solve the singular initial-boundary-value problem that would arise. Thus, some method of averaging must be used to develop a model which captures the finer geometric characteristics of the interface of the media and which is amenable to analysis. For example, in the simplest homogeneous model one introduces the (locally averaged) experimental flow characteristics of the composite medium. For a fluid with *density* u and *pressure* p which is assumed to be slightly compressible, hence, the state equation $u = u_0 \exp(cp) = S(p)$ holds, we have the Darcy law for the *flux*,

$$\vec{J} = -(k/\mu)u\vec{\nabla}p = -(k/c\mu)\vec{\nabla}u \ .$$

Here k is the *permeability* of the medium and μ is *viscosity* of the fluid. The conservation of fluid mass then yields the *classical diffusion equation*

$$\frac{\partial}{\partial t}(mu) - \vec{\nabla} \cdot \frac{k}{c\mu}\vec{\nabla}u = f \tag{1}$$

in which m is effective *porosity* and f is the distributed source.

In order to describe fluid flow through a general heterogeneous medium consisting of two components, such as cells and fissures, one can consider *double porosity* models. The idea is to introduce at each point in space *two* densities u_1, u_2, two pressures p_1, p_2, (etc.), each obtained by averaging in the respective medium over a generic neighborhood sufficiently large to include many cells. There results a pair of "parallel" flow fields governed by their respective equations

$$\frac{\partial}{\partial t}(m_1 u_1) - \vec{\nabla} \cdot \left(\frac{k_1}{c\mu}\vec{\nabla}u_1\right) + q = f$$

$$\frac{\partial}{\partial t}(m_2 u_2) - \vec{\nabla} \cdot \left(\frac{k_2}{c\mu}\vec{\nabla}u_2\right) - q = 0 \ ,$$

where m_j are effective porosities and k_j are permeabilities of the fissure system ($j = 1$) and cell matrix ($j = 2$), as if the interface were impermeable. The term, $q(x,t)$ is the density of the distributed mass flow rate of fluid from the fissures to the cells. When one assumes the flow rate of fluid exchange between the two components is proportional to the product of the pressure difference and the average density over that pressure interval,

$$\rho = \frac{1}{p_2 - p_1}\int_{p_1}^{p_2} S(p)\,dp = \frac{1}{p_2 - p_1}\frac{1}{c}(u_2 - u_1) \ ,$$

to compute the distributed flux, there follows

$$q = \frac{K}{c\mu}(u_1 - u_2)$$

where K is a characteristic of the geometry of the interface in the fissured medium. Thus, with an obvious change of notation we obtain the parabolic system

$$\frac{\partial}{\partial t}(au_1) - \vec{\nabla} \cdot A\vec{\nabla}u_1 + \frac{1}{\delta}(u_1 - u_2) = f \tag{2.a}$$

$$\frac{\partial}{\partial t}(bu_2) - \vec{\nabla} \cdot B\vec{\nabla}u_2 + \frac{1}{\delta}(u_2 - u_1) = 0 \tag{2.b}$$

as the *double porosity parallel model* for flow in a general two-component medium. The parameter $\delta > 0$ can be related to the size of a typical cell and provides a measure of heterogeneity.

The system (2) can be modified to reflect the special characteristics of a fissured medium. The predominant characteristic is that flow proceeds essentially along the fissures: their widths are considerably greater than the pores in the cells, and in addition, when the medium is totally fissured the cells are isolated from one another by the fissures, so there is no direct cell-to-cell flow. In order to model this phenomenon, we set $B = 0$ in (2) and obtain thereby the *first-order kinetic model*

$$\frac{\partial}{\partial t}(au_1) - \vec{\nabla} \cdot (A\vec{\nabla}u_1) + \frac{1}{\delta}(u_1 - u_2) = f \tag{3.a}$$

$$\frac{\partial}{\partial t}(bu_2) + \frac{1}{\delta}(u_2 - u_1) = 0\ . \tag{3.b}$$

Convective or nonlinear terms frequently occur in (3.a), and it is in general a model for mass transport in a porous medium for which (3.b) is a *sorbtion law* which relates the free phase (fissure) concentration to the sorbed (cell) concentration in the medium. Such physically valid sorbtion laws form an important part of the dispersion mechanism. It is generally accepted that the effects of these sorbing sites (or cells) are rather short ranged, but they have been shown repeatedly by experiment to play a significant role. The formation of long tails on the profiles of the free phase concentrations has been experimentally observed, and such tailing effects are recovered from the kinetic model (3) for an *intermediate* range of values for δ.

A second feature of fissured media is that the fissures occupy a much smaller volume than the blocks, so that $a \ll b$ in (3). In particular, the first term, or rate at which fluid is stored in the fissures, is usually not a significant contribution to the equation, so we set $a = 0$ in (3.a) to obtain the *fissured medium system*

$$-\vec{\nabla}\cdot(A\vec{\nabla}u_1) + \frac{1}{\delta}(u_1 - u_2) = f \quad , \quad \frac{\partial}{\partial t}bu_2 + \frac{1}{\delta}(u_2 - u_1) = 0 \ ,$$

as a model of diffusion in which all storage takes place in the cells and all flow takes place in the fissures. By eliminating u_2 we get the equivalent *fissured medium equation*

$$\frac{\partial}{\partial t}b(u_1 - \delta\vec{\nabla}\cdot A\vec{\nabla}u_1 - \delta f) - \vec{\nabla}\cdot(A\vec{\nabla}u_1) = f \ . \tag{4}$$

When $\delta = 0$ this equation reduces to the classical diffusion equation (1). However the solutions of (4) have quite different properties from those of (1).

The double porosity parallel models described above are based on the assumption that the exchange flux, q, has a spatially distributed density proportional to a pressure difference, and they thereby provide a rather simplistic approximation to the true dynamics of the flux exchange. Next we develop a pair of double porosity models which recognize the local geometry of the cell matrix, or a generic *cell model* at each point in the macro-scale of the fissure system and thereby reflect more accurately the flux exchange on the micro-scale of the individual cells. This will be achieved by a mixed coupling whereby fissures affect cells through their boundaries on the micro-scale of the true interface, but cell effects are distributed on the macro-scale of the fissures. This partial-averaging is justified by the smallness of the cells and the higher diffusion rates in the surrounding fissures. One thereby expects an essentially uniform pressure applied by the fissures to the boundary of an individual cell, and it will be necessary to use two spatial scales to implement this construction.

In order to motivate the concept of a *distributed micro-structure* in a very simple setting, we consider a problem from the dynamics of gas absorption. Consider a cylinder of constant cross-section D whose axis is parametrized by $x \in \mathbb{R}$ and which contains an absorbing material. A gas-air mixture is passed at uniform velocity through the cylinder; we shall describe the exchange of gas between this flow field in the pores and the absorbing material fixed in the cylinder. Suppose that at each point $x \in \mathbb{R}$ we are given the cross-section Ω_x of the absorbant; thus $\Omega_x \subset \mathbb{R}^2$ is a domain bounded within D. Let $u(x,t)$ denote the concentration of the gas moving past x in the pores at time $t > 0$ and let $U(x,y,t)$ be the concentration fixed in the absorbent at $y \in \Omega_x$. If we assume the rate of exchange of gas between the absorbent and the pores on the interface $\partial\Omega_x$ is proportional to the concentration difference, then we obtain from conservation-of-mass the dynamical system

$$\frac{\partial}{\partial t}(au) + \frac{\partial}{\partial x}(Au) + \int_{\partial\Omega_x} B\frac{\partial U}{\partial \nu}\, ds = 0\ , \qquad x \in \mathbb{R} \tag{5.a}$$

$$\frac{\partial}{\partial t}(U) = \vec{\nabla}_y \cdot B\vec{\nabla}_y U\ , \qquad y \in \Omega_x\ , \tag{5.b}$$

$$\frac{\partial U}{\partial \nu} + \frac{1}{\delta}(U - u) = 0\ , \qquad s \in \partial\Omega_x\ . \tag{5.c}$$

We have ignored here any diffusion effects along the length of the cylinder in (5.a). The coefficient $a(x)$ represents the area of the effective pores, $D \sim \Omega_x$, the second term is due to the prescribed velocity, and the third term arises from the total flux absorbed out of the pores into the material. Likewise in (5.b) we include only diffusion effects which are transverse to the cylinder and the flux exchanged on the boundary of Ω_x is determined by (5.c). The extensions to include additional diffusion in (5.a) and nonlinear absorption characteristics in (5.c) are evident. The major point in this example is that we have a distributed structure with two scales. The global medium is the cylinder $\mathbb{R} \times D$ which is represented by $\mathbb{R}$, and the concentration u is prescribed there by (5.a). The local dynamics at each point $x \in \mathbb{R}$ is prescribed in the cell Ω_x by (5.b), and these concentrations are coupled on the interface by the flux condition (5.c). The essential concept in this model is that with each point $x \in \mathbb{R}$ we naturally identify a region $\Omega_x \subset \mathbb{R}^2$. This provides our first *distributed micro-structure model*, a continuous distribution over the macro-scale $x \in \mathbb{R}$ of a collection of cells Ω_x and the corresponding micro-scale $y \in \Omega_x$.

Now we describe our first *distributed micro-structure model* of a fissured medium. This consists of a flow region Ω on which is prescribed a continuous distribution of (individually isolated) cells. Thus let Ω be a bounded domain in $\mathbb{R}^n$ with boundary $\Gamma = \partial\Omega$. As before, we let $u(x,t)$ denote the density in the fissure system at each $x \in \Omega$ and time $t > 0$. This fissure density satisfies the diffusion equation

$$\frac{\partial}{\partial t}(a(x)u) - \vec{\nabla}_x \cdot (A(x)\vec{\nabla}_x u) + q(x,t) = f(x,t) , \qquad x \in \Omega \tag{6.a}$$

and appropriate boundary conditions on Γ where $q(x,t)$ is the density of the mass flow of fluid from the fissure system into the cell at x. For each such $x \in \Omega$, let there be given a cell Ω_x, a bounded domain in $\mathbb{R}^n$ with smooth boundary, $\Gamma_x = \partial\Omega_x$. For a function w on Ω_x we denote by $\gamma_x w = w\big|_{\Gamma_x}$ its restriction or *trace* on Γ_x. The fluid density in the cell Ω_x is given by $U(x,z,t)$, $z \in \Omega_x$, and this function satisfies a local diffusion equation

$$\frac{\partial}{\partial t}(b(x)U) - \vec{\nabla}_z \cdot (B(x)\vec{\nabla}_z U) = 0 , \qquad z \in \Omega_x , \tag{6.b}$$

and boundary condition

$$\gamma_x U(x,s,t) = u(x,t) , \qquad s \in \Gamma_x \tag{6.c}$$

for each $x \in \Omega$ and $t > 0$. The total flux that enters the cell is given by

$$q(x,t) = B(x)\int_{\Gamma_x} \frac{\partial U(x,s,t)}{\partial \nu}\, ds = b(x)\frac{\partial}{\partial t}\int_{\Omega_x} U(x,z,t)\, dz \tag{6.d}$$

according to (6.b) and Green's theorem, and this is the distributed sink in (6.a).

The system (6) constitutes our first double porosity model with micro-structure for a fissured medium. The fracture system is a porous medium whose grains are the individual cells Ω_x which are distributed over the region Ω. These blocks do not interact directly with each other. Each is a porous medium whose flow is governed by (6.b) on the micro-scale of Ω_x and whose pressure is matched by (6.c) to that of the surrounding fissures. Finally, the total flux across the block boundary Γ_x given by (6.d) determines a source density for the fissure system given on the macro-scale of Ω by (6.a). We shall refer to the system (6) as the *matched micro-structure model*. Of course it will be supplemented with initial conditions to get a well-posed problem.

The special case of (6) with uniform blocks, *i.e.*, $\Omega_x = \Omega_0$ for all $x \in \Omega$, has an interesting formulation as an implicit convolution evolution equation. In order to obtain a representation of the solution of (6.b) and (6.c), we consider the *cell problem*

$$\frac{\partial}{\partial t}(bV) - \vec{\nabla}_z \cdot (B\vec{\nabla}_z V) = 0 \quad \text{in } \Omega_0$$
$$\gamma_0(V(\cdot,t)) = w(t)\ , \qquad \text{on } \Gamma_0$$

where $w(t)$ is given on $t \geq 0$. The solution is represented by a Green's function as

$$V(z,t) = \int_0^t \int_{\Gamma_0} g(z,s,t-\tau)\, ds\, w(\tau)\, d\tau$$

and, hence, the total flux is

$$q(t) = b\frac{d}{dt}\int_{\Omega_0} V(z,t)\, dz = b\frac{d}{dt}(\tau * w)$$

with the *history function*

$$\tau(t) \equiv \int_{\Omega_0}\int_{\Gamma_0} g(z,s,t)\, ds\, dz\ .$$

Thus the fissure density satisfies the implicit evolution equation

$$\frac{\partial}{\partial t}(a(x)u + b\tau * u) - \vec{\nabla}\cdot(A(x)\vec{\nabla}u) = f(x,t) \ \text{ in } \Omega \tag{7}$$

with a convolution in time. This is the matched model with *uniform micro-structure*. The history function τ displays the response of the generic block to a change in surrounding fissure density. It is singular at $t = 0$ but integrable and completely monotone. Note that (7) agrees with (3) for the very regular $\tau(t) = \frac{1}{b\delta}\exp(-t/b\delta)$, and this is formally close to the classical (1) with $m = a + b$ when $\delta \to 0$ and with $m = a$ when $\delta \to \infty$.

In our second double porosity model of a fissured medium with micro-structure we relax the requirement that the pressures are exactly matched along the interface, and hence that each cell pressure is constant on the boundary, and replace it with the assumption that the cell boundary flux is proportional to the pressure difference on the boundary. As before we have the system of diffusion equations

$$\frac{\partial}{\partial t}(a(x)u) - \vec{\nabla}_x\cdot(A(x)\vec{\nabla}_x u) + q(x,t) = f(x,t)\ , \qquad x \in \Omega \tag{8.a}$$

$$\frac{\partial}{\partial t}(b(x)U) - \vec{\nabla}_z\cdot(B(x)\vec{\nabla}_z U) = 0\ , \qquad z \in \Omega_x\ , \tag{8.b}$$

but they are coupled by the requirements

$$q(x,t) = \frac{1}{\delta}\int_{\Gamma_x} (u(x,t) - \gamma_x U(x,s,t))\, ds\ , \qquad x \in \Omega\ , \tag{8.c}$$

$$B(x)\frac{\partial U(x,s,t)}{\partial \nu} + \frac{1}{\delta}(\gamma_x U(x,s,t) - u(x,t)) = 0\ , \qquad s \in \Gamma_x\ . \tag{8.d}$$

The system (8) is the *regularized micro-structure model.* Note that the uniform case can be formulated as (7). Moreover, this is essentially the system (5) with convection replaced by diffusion.

The micro-structure models incorporate two averaging processes. The pressure in the fissures u was extended to the whole of the domain Ω, and the number of blocks went from a large but finite collection to an uncountable collection of cells, one at each point $x \in \Omega$. We next write down an exact model for the flow in the fissures and blocks, and indicate how this leads intuitively to the matched micro-structure model. Let $\overline{\Omega} = \overline{\Omega}_f \cup \bigcup_{i=1}^N \overline{\widetilde{\Omega}}_i$, where Ω_f represents the fissures, and $\widetilde{\Omega}_i$ are the blocks. The blocks correspond to cells in the micro-model, accordingly we define $\varepsilon_i^n = |\widetilde{\Omega}_i|$ and scale each block by $1/\varepsilon_i$ to give a cell $\Omega_i = \frac{1}{\varepsilon_i}\widetilde{\Omega}_i$. The "exact" equations for flow in the fissures and blocks may then be written as

$$\frac{\partial}{\partial t}(au) - \nabla\cdot(\widetilde{A}\nabla u) = f \ \text{ in } \ \Omega_f \tag{9.a}$$

$$\frac{\partial}{\partial t}(b_i U_i) - \nabla\cdot\left(\frac{1}{\varepsilon_i^2}B_i\nabla U_i\right) = 0 \ \text{ in } \ \Omega_i\ ,\ i = 1,2,\ldots,N \tag{9.b}$$

$$\gamma_i U_i = u \ \text{ and } \ \frac{1}{\varepsilon_i}B_i\frac{\partial U_i}{\partial \nu} = -\widetilde{A}\frac{\partial u}{\partial \nu} \ \text{ on } \ \partial\Omega_i\ ,\ i = 1,\ldots,N\ , \tag{9.c}$$

where we identify $\partial\Omega_i$ with $\partial\widetilde{\Omega}_i$ in (9.c) as above. If the density in the fissures is continuously extended in an arbitrary fashion to all of Ω, an exercise in calculus shows that the system (9) becomes

$$\frac{\partial}{\partial t}(X_{\Omega_f} au) - \nabla\cdot(X_{\Omega_f}\widetilde{A}\nabla u) + q = f \ \text{ in } \ \Omega \tag{10.a}$$

$$\frac{\partial}{\partial t}(b_i U_i) - \nabla\cdot\left[\frac{1}{\varepsilon_i^2}B_i\nabla U_i\right] = 0 \ \text{ in } \ \Omega_i\ ,\ i = 1,2,\ldots,N \tag{10.b}$$

$$\gamma_i U_i = u\big|_{\partial\widetilde{\Omega}_i} \qquad i = 1,2,\ldots,N \tag{10.c}$$

where

$$q(x) = \sum_{i=1}^{N} \frac{1}{|\Omega_i|} \int_{\partial\Omega_i} \frac{1}{\varepsilon_i^2} B_i \frac{\partial U_i}{\partial \nu} \, ds \, \mathbf{X}_i(x) \; , \tag{10.d}$$

X_{Ω_f} is the characteristic function for Ω_f, and $\mathbf{X}_i$ is the characteristic function for Ω_i. The system (10) suggests the micro-structure models, since formally we expect the following limits:

$$\begin{aligned} &X_{\Omega_f} \to m_1 \quad , \quad \text{the fissure volume ratio} \\ &X_{\Omega_f} \widetilde{A} \to A \; , \quad \text{the homogenized coefficients} \\ &\widetilde{\Omega}_i \to x \quad , \quad \text{(for a suitable subsequence)} \\ &\gamma_i \to \gamma_x \quad , \quad U_i(\cdot) \to U(x, \cdot) \\ &b_i \to b(x) \quad , \quad \frac{1}{\varepsilon_i^2} B_i \to B(x) \end{aligned}$$

and

$$q(x) \to \frac{1}{|\Omega_x|} \int_{\partial\Omega_x} B(x) \frac{\partial}{\partial \nu} U(x, \cdot) \; .$$

Substituting the alleged limits into (10) gives the matched micro-structure model (6). The above "limits" suggest that for a given physical problem (with a large but finite number of blocks), an appropriate matched micro-model would have cells whose geometry was similar to that of the blocks, and the cell conductivity should be chosen as $B(x) \approx \frac{1}{\varepsilon_i^2} B_i$ where $\varepsilon_i^n = |\widetilde{\Omega}_i|$ (assuming the cells have unit measure). The block conductivity B_i is expected to be small, as is the block size (and hence ε_i); the formula $B(x) \approx \frac{1}{\varepsilon_i^2} B_i$ shows how these effects balance. If, during the limiting process, $\frac{1}{\varepsilon_i^2} B_i \to \infty$, the block equation will give $U_i = \text{constant}$, so in the limit $u = U$ and we expect to recover the classical model (1). If equation (10.c) is replaced by the mixed condition

$$\frac{1}{\varepsilon_i} B_i \frac{\partial U_i}{\partial \nu} - \frac{\varepsilon_i}{\delta} (\gamma_i U_i - u|_{\partial\widetilde{\Omega}_i}) = 0 \; , \tag{10.c$'$}$$

then the regularized model (8) is obtained. If during the limiting process we assume that $\frac{1}{\varepsilon_i^2} B_i \to \infty$, then the first order kinetic model (3) results. The preceding remarks are intended to merely suggest that micro-structure models can be obtained as limits of exact models. That this is in fact the case and that the "averaged" coefficients in (6) can be given explicitly in terms of those coefficients in the exact model (10) are consequences of homogenization theory.

2. Bibliographical Remarks.

The classical diffusion equation (1) was introduced by J. Fourier at the beginning of the 19th century in his study of heat conduction, and it arose fifty years later in the fundamental work of H. Darcy on diffusion in porous media. During the intervening nearly two centuries this equation has been responsible for the development of major branches of mathematical analysis, and nearly every area of mathematics has been used to some degree to study this very familiar and widely applicable partial differential equation. For a rather exhaustive (and exhausting) bibliography of work on this equation and its applications, we can fortunately refer to the treatise [10].

During the first half of the 20th century there was much interest in various systems of partial differential equations, especially those which arose naturally in applications and could be directly described by the available linear algebra techniques. One finds in the literature of this period many independent works which led to the linear *parabolic system* (2) which naturally occurs in many instances of diffusion of two-component systems [33], [6]. The special case of the first-order kinetic model (3) can be found repeatedly in this period, *e.g.*, [1], and it has been of substantial interest more recently as a model for flow in porous media with adsorption [14], [41], [13], [22], [12] as well as for its nonlinear extensions and the special properties of solutions [16], [19], [37].

This degenerate parabolic system can be written as a *weakly parabolic* equation of viscoelastic type. See [28] for a survey of the relevance of (3) as a diffusion-dispersion model. The fissured medium system or equation (4) dates back to this same period, *e.g.*, [30], and one can consult Chapter 3 of [11] for the rather extensive occurrence of such *pseudo-parabolic* or Sobolev type equations in applications.

Additional and more recent work includes [15], [35], [36], [8], [9], [34].

The concept of the distributed micro-structure model can be found during this period, at least in the technically simpler case of (5) in [29], but the full significance and development of this concept in the form of the systems (6) and (8) was apparently first realized at the middle of this century in the works of [32], [31], [17] from physical chemistry. Problems arising from fluid transport in soils motivated a veritable explosion of interest in such models in the middle of this decade, and much of the work appears to have been developed independently. For example, see [7], [21], [18], [2], [3]. From yet another direction came the application of *homogenization theory* in the work of [40], [25], [26] in which models like (6) are obtained as limits of an *exact* micro-scale model and the effective coefficients are derived explicitly. This technique is described in [24], [27], [5] with a comparison of (3) and (6) and proof of convergence of homogenization is given in [4] for the case of a fissured medium model. A nonlinear parabolic system essentially of the form (8) was directly proposed and studied in [20].

A common feature of all of the preceding works is that they dealt with the case of a *uniform* micro-structure distribution. This minor limitation arose from the use of the functional differential equation (7) as either a vehicle in which to develop the existence-uniqueness theory [2], [27] or a means to construct the model [7], [21]. In fact, the equation (7) is someplaces regarded as the model, and (6) is only an intermediate step toward this goal. As we showed above, the equation (7) can be obtained from (6) or (8) if the cell problem is linear. Then the very general results of [23] can be applied directly to obtain well-posedness results, even for a large class of nonlinear problems in the global medium. The case of a non-uniform distribution of cells, a different cell at each point of the medium, was developed in [38] by considering the systems (6) and (8) directly on a large product of spaces rather than (7) on a simpler space. A more significant advantage of this approach is that the full parabolic regularizing effects and related estimates could be obtained, properties that can not be obtained from (7). Moreover, it was shown in [38] that the solutions of (3) and (8) converge to those of (1) and (6), respectively, as $\delta \to 0$; in fact, (8) was used there as a "penalty approximation" of the "constrained" problem (6). Also, it was shown there that (6) and (8) converge similarly to (1) and (3) as the coefficient $B(x) \to +\infty$. These results establish the "formally" obvious but technically important connections between these model problems. Similar systems with multiple nonlinearities are being developed [39] as well as those in which convection dominates the global flow field.

References

[1] A. Anzelius, *Über Erwärmung vermittels durchströmender Medien*, Zeit. Ang. Math. Mech., **6** (1926), 291–294.

[2] T. Arbogast, *Analysis of the simulation of single phase flow through a naturally fractured reservoir*, SIAM J. Numer. Anal., **26** (1989), 12–29.

[3] T. Arbogast, *The double porosity model for single phase flow in naturally fractured reservoirs*, Inst. Math. Appl., **295** (1987).

[4] T. Arbogast, J. Douglas, U. Hornung, *Derivation of the double porosity model of single phase flow via homogenization theory*, SIAM J. Math. Anal., **21** (1990), 823–836.

[5] T. Arbogast, J. Douglas, U. Hornung, *Modeling of naturally fractured reservoirs by formal homogenization techniques*, to appear.

[6] G.I. Barenblatt, I.P. Zheltov, I.N. Kochina, *Basic concepts in the theory of seepage of homogeneous liquids in fissured rocks*, J. Appl. Math. and Mech., **24** (1960), 1286–1303.

[7] J.A. Barker, *Block-geometry functions characterizing transport in densely fissured media*, J. of Hydrology, **77** (1985), 263–279.

[8] M. Böhm and R.E. Showalter, *Diffusion in fissured media*, SIAM J. Math. Anal., **16** (1985), 500–509.

[9] M. Böhm and R.E. Showalter, *A nonlinear pseudoparabolic diffusion equation*, Siam J. Math. Anal., **16** (1985), 980–999.

[10] J.R. Cannon, *The One-Dimensional Heat Equation*, in "Encyclopedia of Mathematics and its Applications", G.-C. Rota, ed., Addison-Wesley, Reading, Mass., 1984.

[11] R.W. Carroll and R.E. Showalter, "Singular and Degenerate Cauchy Problems", Academic Press, New York, 1976.

[12] E. Charlaix, J.P. Hulin, T.J. Plona, *Experimental study of tracer disperion in sintered glass porous materials of variable compation*, Phys. Fluids, **30** (1987), 1690–1698.

[13] K.H. Coats, B.D. Smith, *Dead-end pore volume and dispersion in porous media*, Trans. Soc. Petr. Eng., **231** (1964), 73–84.

[14] H.A. Deans, *A mathematical model for dispersion in the direction of flow in porous media*, Trans. Soc. Petr. Eng., **228** (1963), 49–52.

[15] E. DiBenedetto and R.E. Showalter, *A pseudo-parabolic variational inequality and Stefan problem*, J. Nonlinear Anal., **6** (1982), 279–291.

[16] E. DiBenedetto and R.E. Showalter, *A free-boundary problem for a degenerate parabolic system*, Jour. Diff. Eqn., **50** (1983), 1–19.

[17] P.F. Diesler, Jr. and R.H. Wilhelm, *Diffusion in beds of porous solids: measurement by frequency response techniques*, Ind. Eng. Chem. **45** (1953), 1219–1227.

[18] J. Douglas, P.J. Paes Leme, T. Arbogast, T. Schmitt, *Simulation of flow in naturally fractured reservoirs*, in Proceedings, Ninth SPE Symposium on Reservoir Simulation, Society of Petroleum Engineers, Dallas, Texas, 1987, pp.271–279, Paper SPE 16019.

[19] C.J. van Duijn and P. Knabner, *Solute transport through porous media with slow adsorption*, to appear.

[20] A. Friedman, A. Tzavaras, *A quasilinear parabolic system arising in modelling of catalytic reactors*, Jour. Diff. Eqns., **70** (1987), 167–196.

[21] M.Th. van Genuchten, F.N. Dalton, *Models for simulating salt movement in aggregated field soils*, Geoderma, **38** (1986), 165–183.

[22] M.Th. van Genuchten, P.J. Wierenga, *Mass transfer studies in sorbing porous media I. Analytical Solutions*, Soil Sci. Soc. Amer. J., **40** (1976), 473–480.

[23] G. Gripenberg, *Volterra integro-differential equations with accretive nonlinearity*, Journ. Diff. Eqns., **60** (1985), 57–79.

[24] U. Hornung, *Miscible displacement in porous media influenced by mobile and immobile water*, Rocky Mountais J. Mathem., to appear.

[25] U. Hornung and W. Jäger, *A model for chemical reactions in porous media*, in "Complex Chemical Reactions, Modeling and Simulation" (J. Warnatz, W. Jäger, eds.), Chemical Physics, Springer, Berlin, 1987.

[26] U. Hornung and W. Jäger, *Diffusion, convection, adsorption, and reaction of chemicals in porous media*, J. Differ. Equat., to appear.

[27] U. Hornung and R.E. Showalter, *Diffusion models for fractured media*, Jour. Math. Anal. Appl., **147** (1990), 69–80.

[28] F.T. Lindstrom, M.N.L. Narasimham, *Mathematical theory of a kinetic model for dispersion of previously distributed chemicals in a sorbing porous medium*, SIAM J. Appl. Math., **24** (1973), 496–510.

[29] A.N. Lowan, *On the problem of the heat recuperator*, Phil. Mag. **17** (1934), 914–933.

[30] E.A. Milne, *The diffusion of imprisoned radiation through a gas*, Jour. London Math. Soc., **1** (1926), 40–51.

[31] J.B. Rosen, *Kinetics of a fixed bed system for solid diffusion into spherical particles*, J. Chem. Phys., **20** (1952), 387–394.

[32] J.B. Rosen and W.E. Winshe, *The admittance concept in the kinetics of chromatography*, J. Chem. Phys., **18** (1950), 1587–1592.

[33] L.I. Rubinstein, *On the problem of the process of propagation of heat in heterogeneous media*, Izv. Akad. Nauk SSSR, Ser. Geogr., **1** (1948).

[34] J. Rulla and R.E. Showalter, *Diffusion with prescribed convection in fissured media*, Diff. and Integral Equations, **1** (1988), 315–325.

[35] R.E. Showalter, *Local regularity, boundary values and maximum principles for pseudoparabolic equations*, Applicable Analysis, **16** (1983), 235–241.

[36] R.E. Showalter, *The fissured medium equation*, in "Physical Mathematics and Nonlinear Partial Differential Equations," J.H. Lightbourne, III and S.M. Rankin, III, eds., Marcel Dekker, New York, 1985.

[37] R.E. Showalter and N.J. Walkington, *A diffusion system for fluid in fractured media*, Differential and Integral Eqns., **3** (1990), 219–236.

[38] R.E. Showalter and N.J. Walkington, *Micro-structure models of diffusion in fissured media*, Jour. Math. Anal. Appl., to appear.

[39] R.E. Showalter and N.J. Walkington, *Diffusion of fluid in a fissured medium with micro-structure*, to appear.

[40] Ch. Vogt, *A homogenization theorem leading to a Volterra integro-differential equation for permeation chromotography*, preprint #155, SFB **123**, Heidelberg, 1982.

[41] J.E. Warren, P.J. Root, *The behavior of naturally fractured reservoirs*, Soc. Petr. Eng. J., **3** (1963), 245–255.

Transport in Porous Media 6: 581–606, 1991
© 1991 *Kluwer Academic Publishers. Printed in the Netherlands.*

Review Article:

CHARACTERIZATION OF POROUS MEDIA - PORE LEVEL

F.A.L. DULLIEN
Department of Chemical Engineering
University of Waterloo
Waterloo, Ontario N2L 3G1
Canada

ABSTRACT. Pore size and pore size distribution are defined, 1-D pore structure models, 2-D and 3-D network models of pore structure are reviewed. Simulation of capillary pressure and relative permeability curves with the help of network models of pore structure is discussed. Pore structure determination from serial sections is outlined. Phase immobilization technique is reviewed.

Key words. Network models, pore structure, phase immobilization.

1. INTRODUCTION - BACKGROUND

Characterization of porous media on the pore (microscopic) level is done for the purpose of understanding, modeling, and in some cases, controlling the behavior and the properties of the macroscopic medium. The concept of what is 'macroscopic' is an intuitive one and it usually means a sample of the medium on which tests are performed in the lab. As network modeling of transport properties [21] has indicated that a cubic network of 15 x 15 x 15 was sufficiently large to predict the same behavior as larger networks, this may be taken as an indication of what constitutes the minimum size of a homogeneous macroscopic system in terms of nodes, representing pore bodies. When referring to macroscopic properties measured in the lab it is also implied that the sample used was 'homogeneous', i.e., if the sample were broken up into two or more pieces, each of which is still macroscopic, each piece would exhibit the same value of the macroscopic property as the original (unbroken) sample. As the sample may be more nearly homogeneous with respect to one property (e.g. the porosity) than another property (e.g. the permeability), a truly homogeneous sample of a porous medium is required to be homogeneous with respect to all the properties which are related to the pore structure. In network modeling 'homogeneity' is guaranteed by distributing the different pore sizes of a given probability density function (pdf) over a network of uniform structure in space in such a way that there is the same probability for each node (or bond) of the network to be assigned a particular pore size, independently of the position of the node (or bond) in the network. As a result, the pore sizes at any node (or bond) are independent of each other, i.e., there is no spatial structure of the distribution of pore sizes. The 'homogeneity' defined in this manner is a very special case of a much broader class of 'homogeneity' in the stochastic sense of the word which permits a 'spatial structure' of the property and stipulates only 'ergodicity' and 'stationarity' of the property, implying the same pdf of all realizations of the property and the same mean value of the property calculated at any point in the medium. As the scale of investigation (the size of the 'sample') increases it is

usually found that spatial structure becomes very important and homogeneity in the narrow, laboratory sense does not exist even approximately.
It is generally realized that it is possible to model transport properties of porous media without any reference to pore geometry, and merely use a large number of adjustable parameters in the model that do not have any physical meaning. In this author's opinion, models of this kind are less useful in facilitating our understanding of the observed physical phenomena than those which incorporate the basic features of pore morphology. This point of view finds ample support in the excellent critical studies published by van Brakel [59] and van Brakel and Heertjes [60]. Pore structure is preferably determined on the basis of visualization of the pore space, using one of several available techniques, followed by computerized reconstruction of the essential features of pore geometry in 3-D.

The main transport properties a good pore structure model should simulate with the same values of the parameters include the effective molecular diffusion coefficient, the absolute permeability, the dispersion coefficient (function of Peclet number), drainage and imbibition capillary pressure curves, dendritic portion of the nonflowing parts of saturations, saturation vs. height of capillary rise, rate of capillary rise, relative permeabilities vs. saturation (the last five are also contact angle and history dependent), formation factor, resistivity index and drying.

1.1. Porosity

The term "pore structure" means different things to different people. On the lowest level of information it is simply interpreted as the "porosity" ϕ, i.e., the volume fraction of pore space in the porous material. The basis of this is the belief that other properties such as e.g. the permeability, k, defined by Darcy's law

$$Q = \frac{k}{\mu} A \frac{\Delta p}{\ell} \tag{1}$$

where Q is the volume flow rate, μ the viscosity of the fluid, A the cross sectional area of the one dimensional sample of length ℓ and Δp the pressure drop, can be predicted from the porosity.

1.2. Pore Size Distribution - Experiment

On a somewhat more sophisticated level of knowledge pore structure is interpreted as a characteristic "pore size", which is sometimes also called "porosity". Most generally, however, "pore structure" is identified with a so-called "pore size distribution", characteristic of the sample of the porous material. "Pore size distribution" is a poorly defined quantity, partly because it depends, sometimes very markedly, on the particular method used in its determination. The general procedure used for the determination of a pore size distribution consists of measuring some physical quantity in dependence on another physical parameter under the control of the operator and varied in the experiment. For example, in mercury porosimetry, the volume of mercury penetrating the sample is measured as a function of the pressure imposed on the mercury, in vapor sorption, the volume of gas absorbed is measured as a function of the gas pressure, the volume of liquid displaced miscibly is measured as a function of the volume of displacing liquid injected into the sample in a miscible displacement experiment, etc.

1.3. 1-D Pore Structure Models

The experimental data have invariably been interpreted in terms of an arbitrary model of pore structure, the most popular one consisting of a bundle of parallel capillary tubes of equal length and distributed diameters. The fact that this model may give rise to vastly different "pore size distributions" when used in conjunction with the results obtained in different types of experiments on the same sample has been demonstrated by Klinkenberg [1] for the case of mercury porosimetry and miscible displacement, as illustrated in Figure 1.

1.4. Dead-end Pores and Periodically Constricted Tubes

More sophisticated 1-D models have included dead-end pores, called also "ink-bottle" pores, "pockets" or "Turner structures", depending on their shape and also periodic variations of the diameter of each capillary tube, the so-called "periodically constricted tube" models. Both types of pore models introduce the concept of "pore throats", the local minima in the pore size that separate "pore bodies". These models have been reviewed in [2]. Each is designed to account only for certain properties of the porous medium under study while quite unable to account for other properties. In order to account for 3-D flow, intrinsically 1-D models have often been generalized in three dimensions [2].

1.5. Definition of Pore Size Distribution

The definition of pore size distribution, in the usual sense, is "the density function giving the distribution of pore volume by a characteristic pore size." Whereas the volume parameter is usually measured directly, the characteristic pore size is always calculated from some measured physical parameter in terms of the arbitrary model of pore structure. Owing to the complexities of pore geometry, the characteristic pore size is often not at all characteristic of the pore volume to which it has been assigned. In mercury intrusion porosimetry, for example, the volume of the pore space penetrated through a pore throat is assigned to the size of the throat, resulting in an unrealistic picture of the real pore structure, as shown in Figure 2. The realization of this state of affairs has given rise to the custom of referring to "entry pore size."

Very recently, constant rate injection mercury porosimetry has been used to extract more information on pore structure from the fluctuations observed in the porosimetry trace representing abrupt pressure changes related to the jumps of mercury from pore throats into (larger) pore bodies (Haines jumps) with concomitant redistribution of mercury also at other locations [57]. The lower extremes of the fluctuating pressure are related to the size of pore bodies. This method has promise to develop into a nondestructive quantitative technique for the determination of both pore throat and pore body size distributions.

1.6. Pore Size

Another matter that needs clarification is the definition of "pore size." Only if the pores were cylindrical tubes of uniform diameter, or spherical bodies, would the pore size be unique. As neither is the case, "pore size" needs to be defined. A convenient definition is twice the hydraulic radius, $2r_H$, which is either identical or very close to the mean radius of curvature

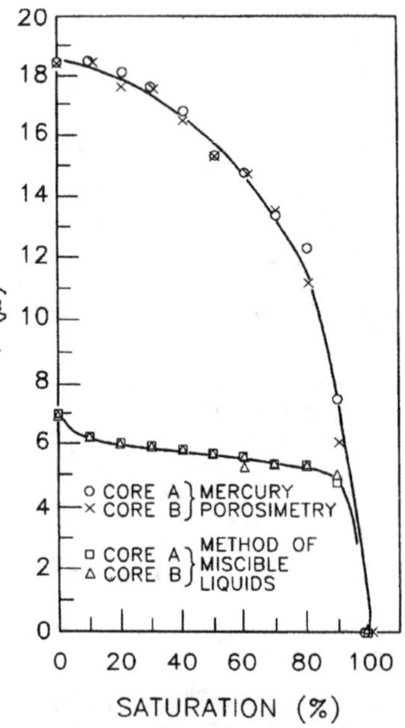

Figure 1. Cumulative pore size distribution curves of Bentheim sandstone. Permeability: 975×10^{-11} cm^2. Porosity: 0.289 (after Klinkenberg [1]).

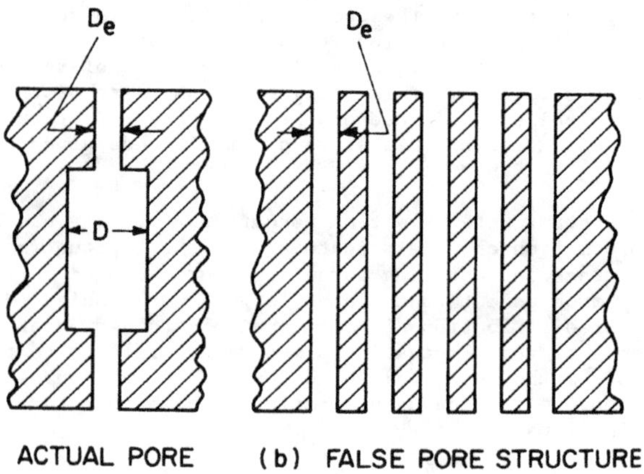

Figure 2. Illustration of error introduced by the usual interpretation of mercury intrusion porosimetry data: (a) actual pore; (b) false interpretation of data. (D_s = entry pore throat diameter; D_ℓ = pore body diameter) (after Dullien [58]).

$$\frac{1}{r_m} = \frac{1}{2}\left(\frac{1}{r_1} + \frac{1}{r_2}\right) \tag{2}$$

of the interface separating two immiscible fluids in mechanical equilibrium in the pore, for the special case of zero contact angle (see Table 1 [3]). The mean radius of curvature, r_m, can be calculated from measured capillary pressures, P_c, by Laplace's equation of capillarity

$$P_c = \frac{2\sigma}{r_m} \tag{3}$$

where σ is the interfacial tension. For the case of nonzero contact angle θ the values of the pore radius r_m in Table 1 must be replaced by R, according to the relation,

$$R = r_m \cos\theta \quad . \tag{4}$$

The definition of hydraulic radius r_H of a capillary of uniform cross-section is

$$r_H = \frac{\text{volume of capillary}}{\text{surface area of capillary}} \tag{5}$$

For the case of a variable cross-section the above definition can be generalized for any normal cross-section of the capillary as follows:

$$r_H = \frac{\text{area of cross section}}{\text{length of perimeter of cross section}} \tag{6}$$

The applicability of Equation (6), however, is limited to the special case of capillaries of a rotational (axial) symmetry that have normal cross-section. For the general case of irregular capillaries the minimum

TABLE 1. List of Comparative Values to Show Equivalence of the Reciprocal Hydraulic Radius ($1/r_H$) and Twice the Reciprocal Mean Radius of Curvature $2/r_m = [(1/r_1) + (1/r_2)]$ in a Capillary[a] (after Carman [3]).

Cross-section	$(1/r_1) + (1/r_2)$	$1/r_H$
Circle	2/r	2/r
Parallel plates	1/b	1/b
a:b = 2:1	1.50/b	1.54/b
Ellipse a:b = 5:1	1.20/b	1.34/b
a:b = 10:1	1.10/b	1.30/b
Rectangle	1/a + 1/b	1/a + 1/b
Equilateral triangle	$2/r_i$	$2/r_i$
Square	$2/r^i$	$2/r_i$

[a] r_i is the radius of the inscribed circle.

value of the ratio given by the right hand side of Equation (6) must be found by varying the orientation of the sectioning plane about the same fixed point inside the capillary. The minimum value of this ratio is, by definition, the hydraulic radius r_H of the irregular capillary at the fixed point. The value of r_H of a section is assigned to the center of gravity of the section. Both definitions (i.e., r_m and r_H) are best suited to the case of pore throats that control both capillary penetration by a nonwetting fluid into the porous medium and the flow rate of fluids through the porous medium. The size of a pore body is not readily related in a unique manner to any measurable physical quantity and the problem of characterizing the size of a pore body is best dealt with by using photomicrographs of sections made through the porous medium where the pore body is made visible. The problem of defining the size of a pore body is similar to the problem of defining the size of an irregularly shaped particle.

2. NETWORK MODELS OF PORE STRUCTURE

2.1. 2-D Network Models

The most fundamental flaw of all the simple models of pore structure is that they do not account for the fact that in permeable porous media all the conducting pores are interconnected and form a continuum of a network of pores. In a network where pores of different sizes are interconnected either in a random or in a correlated manner there is a large number of different pathways characterized by different resistances to transport. In any given porous medium the distribution of paths of different resistances depends on the nature of the particular transport phenomenon and, as a result, a variety of phenomena may occur in 2-D and 3-D networks that are impossible in 1-D models. Repeating an intrinsically 1-D model in the three spatial directions, even if tubes are made to intersect, cannot account for these phenomena [2]. Even though there are important differences between the properties of 2-D and 3-D networks, nevertheless the first move in the direction of introducing accurate pore structure models was the pioneering work by Fatt [4,5,6] who proposed random 2-D network models of pore structure for the first time. Whereas Fatt was primarily interested in immiscible displacement, Simon and Kelsey [7,8] used 2-D network models for the simulation of miscible displacement. These studies, important as they were owing to their pioneering character, were only of qualitative nature because

i. the networks were 2-D,
ii. the size of the network was too small,
iii. the networks were regular,
iv. the networks consisted only of tubes, in analogy with a network of resistors,
v. the geometry of the tubes and the distribution of the geometry over the network ("size distribution") did not correspond to that of any real porous medium because this was not available.

2.2. Percolation Theory - 3-D Network Models

In the same time period the powerful mathematical theory of percolation was developed (e.g., [9,10,11]) that was also suitable for the rigorous treatment of immiscible displacement and two-phase flow phenomena in pore networks of infinite size.

The first published reference in which pore structure was modelled by both 2-D and 3-D networks appears to be the work of Chatzis and Dullien [12], which was followed by a vastly improved treatment by the same two authors a year later [13]. The unique features of this treatment, the final form of which was published in the English language literature only in 1985 [14], included 3-D network models of pore structure consisting of pore bodies situated at the nodes of the network (sites) connected by pore throats, modelled by the bonds of the network. The two 3-D networks used, i.e., simple cubic and tetrahedral networks of coordination numbers 6 and 4, respectively, gave similar results.

Modeling was carried out in three distinct stages. In the first stage integers ranging from 1 to 100 were assigned to the sites merely to indicate the hierarchy of pore sizes, e.g. 55 > 22, and not to give the actual size of any pore. The first stage was sufficient and necessary to simulate the penetration pattern of the network. It was assumed that a bond can never be larger than either of the two sites connected by it. Hence the same integer characterizing the smaller one of the two sites was automatically also assigned to the bond between the two sites. This resulted in bond sizes correlated with the site sizes ("bond-correlated site percolation").

In the second stage, numbers representing the actual pore sizes were assigned to the sites and the bonds. Thereby equilibrium capillary pressures could be assigned to the pores controlling penetration at each point of the process. In the final stage, volumes were assigned to each pore thus making possible the calculation of saturations at each point of the process. The three stages of modeling are outlined next in some detail.

The penetration of the network by a nonwetting fluid (drainage) was simulated by the following algorithm. Initially all the sites and bonds were assumed "closed." The algorithm was started by first declaring the largest sites "open", then the second largest sites, and so forth. A bond became open if and only if both sites at its two ends were "open." "Open" sites and bonds communicating with the face of the network exposed to the fluid were automatically penetrated. The rest of the faces of the network were assumed impervious. The fractional numbers of sites and bonds that were penetrated was recorded as functions of the fractional number of "open" sites. The fractional number of "open" sites, p_s, at which the penetration first reaches the opposite face of the network is called the "breakthrough" value. In the case of a network of infinite size this value is the "critical percolation probability" or "percolation threshold" which is known from percolation theory. Close agreement with the published percolation theory value has been obtained by repeating the game in an 18 x 18 x 12 mesh size cubic network and taking the average breakthrough value. The process is continued until all sites and bonds are penetrated. The fractional number, or probability, of "open" bonds, p_b, is related to the fractional number, or probability, of "open" sites, p_s, as follows

$$p_b = p_s^2 \tag{7}$$

owing to the assumption that both sites at the two ends of a bond must be "open" for a bond to be also "open." As a result, relationships have been obtained for the fractional numbers of penetrated sites and bonds as functions of the fraction of open bonds, p_b. These relationships were used for predicting the mercury porosimetry curves of sandstone samples, among other things. Before one can attempt to do this, first pore sizes and volumes must be assigned to the sites and the bonds. The pore diameters, D_b for the bonds and D_s for the sites, are related to p_b and p_s, respectively, by the following relationships

$$p_b = \int_{D_b}^{D_{bmax}} f_b(D_b)\, d\, D_b \tag{8}$$

and

$$p_s = \int_{D_s}^{D_{smax}} f_s(D_s)\, d\, D_s \tag{9}$$

where $f_b(D_b)$ and $f_s(D_s)$ are the bond (pore throat) and site (pore body) diameter distribution densities. Equations (8) and (9) follow from the definitions of the quantities appearing in them and are, in effect, identities. They were solved for D_b and D_s, using $f_b(D_b)$, $f_s(D_s)$, p_b and p_s. The ranges of pore sizes used were consistent with photomicrographs prepared of polished sections of sandstone samples that had been previously saturated with Wood's metal, but the functions, shown in Figure 3, were adjusted so as to obtain the best agreement between prediction and experiment for the case of a particular Berea sandstone sample.

For each value of D_b the capillary pressure of mercury penetration, P_c, was calculated by the relation

$$P_c = \frac{4\sigma \cos \theta}{D_b} \tag{10}$$

Based on petrographic studies [15] the pores in sandstones are slit-shaped. Hence, geometrically similar, slit-shaped pore bodies and pore throats have been assumed, where the pore "diameter" (D_s or D_b) is the width of the slit and the other two dimensions of the slit are $L_1 = C_1(D)^{1/2}$ and $L_2 = C_2(D)^{1/2}$, resulting in the following expressions for the volume of a pore body V_s and a pore throat V_b, respectively

$$V_s = D_s L_{1s} L_{2s} = D_s^2 C_{1s} C_{2s} = D_s^2 . \ell_s \tag{11}$$

and

$$V_b = D_b L_{1b} L_{2b} = D_b^2 C_{1b} C_{2b} = D_b^2 . \ell_b \tag{12}$$

where ℓ_s and ℓ_b are dimensional constants.

Thus the predicted volume of pores penetrated by the mercury at any given capillary pressure could be readily calculated. The assumed fraction of volume contributed by the pore throats, anywhere in the range of under 50%, had very little effect on the total volume predicted. An additional fine point considered in the calculation is that in the case of capillaries of

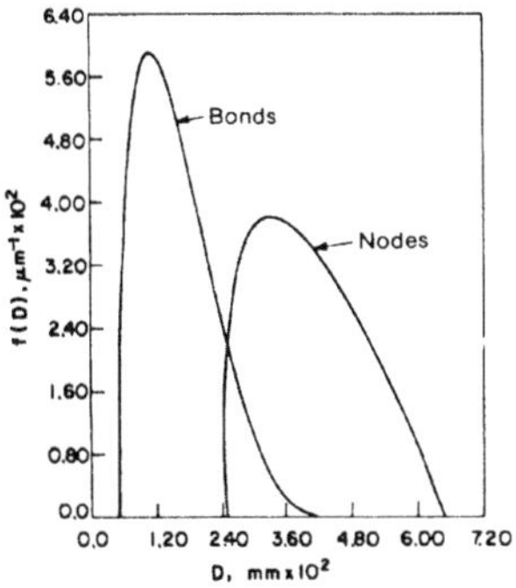

Figure 3. Density distributions of pore throat (= bond) and pore body (= node) diameters (after Chatzis and Dullien [14]).

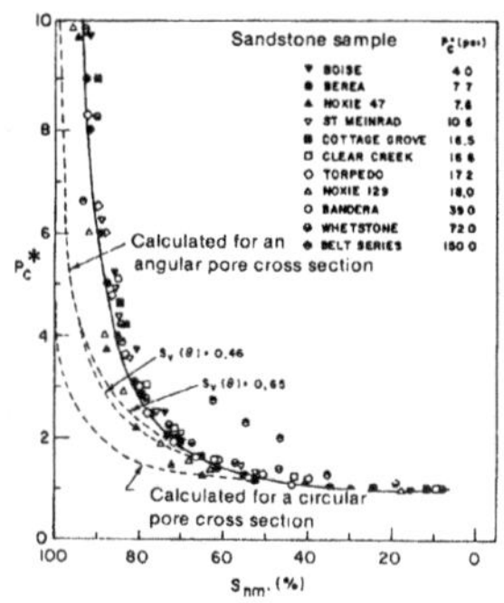

Figure 4. Dimensionless mercury intrusion porosimetry curve of sandstone samples (after Chatzis and Dullien [14]).

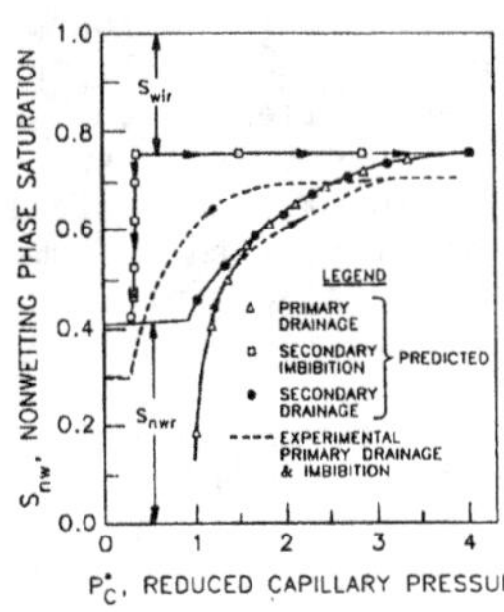

Figure 5. Simulated and experimental Berea sandstone capillary pressure curves (after Diaz et al. [16]).

irregular, rather than circular, cross-section penetration of the capillary by mercury does not result in complete filling of the entire cross-section because it would take a higher capillary pressure for the mercury to penetrate the corners than the pressure required to penetrate the core of the pore. Three calculations were carried out: one for circular pore cross section, one for the case of 0.46 and another for the case of 0.65 unfilled volume fraction when the pore was first penetrated by mercury. As the capillary pressure on the mercury is increased progressively these empty fractions keep decreasing proportionately to the square of the capillary pressure (i.e., two-fold increase in the capillary pressure results in a four-fold decrease in the unfilled volume).

The results of these calculations have been compared with experimental mercury intrusion porosimetry curves obtained on ten different sandstone samples of permeabilities ranging from 5370 md to 0.36 md. The data have been plotted in reduced form in terms of $P_c^* = P_c/P_c^\circ$, where P_c° is the breakthrough capillary pressure of the particular sample, vs. mercury saturation (S_{nm}%) in Figure 4. The fact that, with the exception of Belt Series, the data lie on the same curve indicates that the pore structures of the remaining nine sandstones are similar. The calculated curve starts predicting unrealistically high mercury saturations at higher saturations. The assumption of partially unfilled pores does not eliminate this discrepancy completely which remains in the range of 5-10% pore volume. The most likely explanation for this is the presence of micropores in the cementing materials (clays) of the sandstones which were not penetrated by mercury under the capillary pressures used in the experiments and which were not included into the network model.

The same network model was used subsequently to model both the drainage and the imbibition oil-water capillary pressure curves in sandstones [16] where also the trapping of each phase by the other was simulated. The experimental primary drainage curve has been predicted satisfactorily. In this case, too, lower residual wetting phase (water) saturation was predicted than the measured value (see Figure 5). The reason for this discrepancy is probably the water present in the micropores of the cementing materials (clays) the existence of which was not taken into account in the network model. The experimental secondary imbibition curve, however, indicates a far more gradual displacement of the non-wetting phase (oil) than the predicted trend. The reason for this is that in the simulation of imbibition the sites of the network, representing pore bodies, were assumed to control the displacement process. Technically speaking, in the simulation of imbibition first the smallest sites were allowed to be "open", then the second smallest sites, and so forth. Any "open" site in communication with the wetting fluid was automatically penetrated. The bond was also automatically penetrated along with the smaller one of the two sites connected by it. As the imbibition capillary pressure was calculated on the basis of capillary equilibrium in a pore body (site), the value obtained for breakthrough of the wetting phase turned out to be very small and after breakthrough the imbibition process was completed over a very narrow range of capillary pressures. Visual observations of imbibition in transparent capillary micromodels and other experiments have shown that, imbibition is not controlled by the pore bodies because the imbibing fluid does not always advance in a pistonlike manner [17,18,19,20,21]. Instead, it often propagates in pore edges, wedges, corners and surface grooves and it can pass through pore bodies while filling them only fractionally. As a result, the wetting fluid does not have to fill relatively large pore bodies in its path before it can fill pore throats and relatively small pore bodies over the entire network.

In addition to simulating the capillary pressure curves, the information obtained from the cubic network model on the accessibility and occupancy of the pores of the network by the wetting and the non-wetting fluids, respectively, has been used also for predicting the relative permeability curves [14,22,23].

Relative permeabilities k_{ri} are defined by an extension of Darcy's law [Equation (1)] to every fluid i in two- or three-phase flow

$$Q_i = \frac{k_{ri} k}{\mu_i} A \frac{\Delta p_i}{\ell} \tag{1a}$$

where k_{ri} are assumed to be functions of the saturation only. Thereby the assumption of uncoupled flow of the fluids is implied which does not always apply [62]. The same choice of density distributions of pore throat and pore body diameters which were used in the simulation of the capillary pressure curves resulted in excellent predictions of the relative permeability curves in drainage. The imbibition-type relative permeability curves are in better agreement with experiment than the corresponding capillary pressure curves because the capillary pressure is absent from relative permeability curves (see Figure 6) for the same reasons already stated under the discussion of simulation of capillary pressure curves.

The above contributions, discussed in some detail, are distinguished from the rest of the literature of network modelling and percolation theory by both the application of consistent photomicrographic pore structure data in the model calculations and the regular comparisons made between prediction and experiment on specific samples of porous media.

An incomplete list of references to (largely theoretical) contributions to the field of application of percolation theory and network modelling to porous media is [25,26,27,28,29,30,31,32,33,34,35, 36,37]. An interesting computerized network simulation study of drainage and imbibition capillary pressure curves in dependence of a variety of pore structure parameters has been published by Li et al. [38]. The stated purpose of this work has been to interpret the pore structure of porous media from the capillary pressure curves and scanning loops. While a number of trends have been observed in this work, the authors did not work out a procedure leading to the determination of the density function of pore throat and pore body diameters, coordination numbers, the nature of correlation between neighboring pore body and pore throat sizes and the type of pore shapes present in a particular sample. Another recent important contribution is by Lenormand et al. [39] involving both experiments and numerical modelling of immiscible displacement in networks. In contrast to [38], in this work no attention is paid to the effects of pore structure on the immiscible displacement process but instead the emphasis was on the role played by the capillary number and the viscosity ratio in determining whether there is (a) stable displacement, (b) viscous fingering or (c) capillary fingering.

There are at the present powerful network simulation techniques available to predict capillary pressure curves, relative permeabilities, immiscible displacement and also miscible displacement (hydrodynamic dispersion) — the

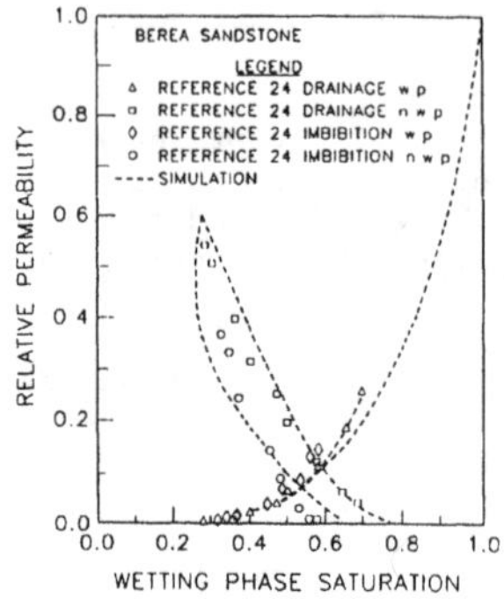

Figure 6. Simulated and experimental oil-water relative permeability curves for Berea sandstones (after Kantzas and Chatzis [22]).

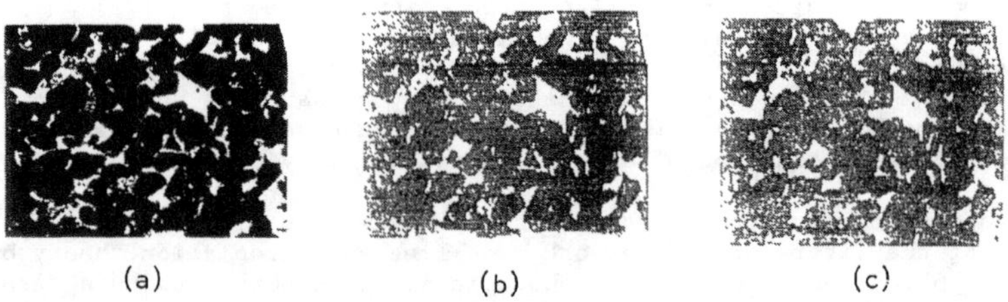

Figure 7. Representative serial section: (a) photomicrograph; (b) original digitized picture; (c) digitized picture after filtering to remove non-connected features (after Kwiecien et al. [55]).

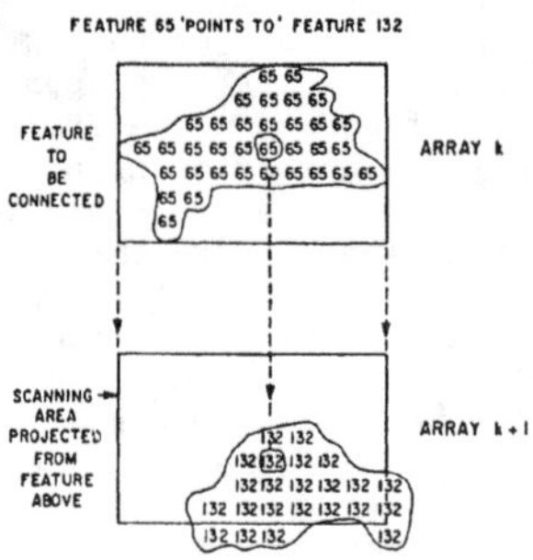

Figure 8. Connection of features on adjacent serial sections by the overlap criterion (after Macdonald et al. [48]).

latter not reviewed here — for the case of arbitrarily assumed pore structures, modelled by the network. It is reasonable to expect that these simulators would be able to predict the actual behavior in samples of real porous media if the correct pore structure data of the samples were used as input. It is, therefore, of considerable practical importance to develop techniques to obtain such data. In the remaining part of this review attempts to develop a technique for the determination of the accurate pore structure of samples of real porous media are presented and discussed. All of these attempts are based on computer reconstruction of the pore structure from serial sections of the porous sample.

3. PORE STRUCTURE DETERMINATION FROM SERIAL SECTIONS

Serial sectioning of samples has been performed for some time by metallographers for the purpose of determining the genus of a phase [40,41,42,43]. More recently this technique has been also applied to porous media [44,45,46,47,48]. The data used in [48] has formed the basis of a number of pore structure studies. The data were generated by first injecting, under high pressure, molten Wood's metal into a previously evacuated sandstone sample. After solidification of the metal the rock matrix was dissolved with hydrofluoric acid and replaced with clear epoxy resin. A piece of this sample of dimensions 1310 x 1040 x 762 μm was encased in epoxy resin and its surface was successively ground and polished to remove layers of about 10 μm thickness. 78 layers were removed and each layer was photographed. A sample photomicrograph is shown in Figure 7. The white features represent the pore space as the metal is reflective and the black regions represent the rock matrix since the epoxy is clear.

3.1. Determination of Genus

The micrographs were entered into the computer where they were digitized then connection between features in neighboring photomicrographs was established based on the assumption that two features are connected if and only if they overlap. Connection of two pixels between two features present in two neighboring micrographs k and k + 1 is illustrated in Figure 8. For the purpose of topological studies, where distances and directions play no role, this kind of a treatment of the data results in a branch-node chart of the sample from which the genus may be readily determined. The genus is defined as the number of non-intersecting cuts which can be made upon a surface without separating it into disconnected parts. It has been shown [49] that the genus of the enclosing surface is numerically equal to the connectivity of the branch-node network derived from that surface. The connectivity is a measure of the number of independent paths between two points in the pore space and, hence, of the degree of interconnectedness of the pores. The genus G is given by

$$G = b - n + N \tag{13}$$

where b is the number of branches, n is the number of nodes and N is the number of separate networks in the sample. Simple examples for the calculation of genus are shown in Figure 9. A small portion of a serially sectioned sample is shown in Figure 10(a). In Figure 10(b) the corresponding branch-node chart is shown. The surface nodes introduce complications because it is not known how they are connected on the outside of the sample. The surface nodes in the plane of polish are visible, but

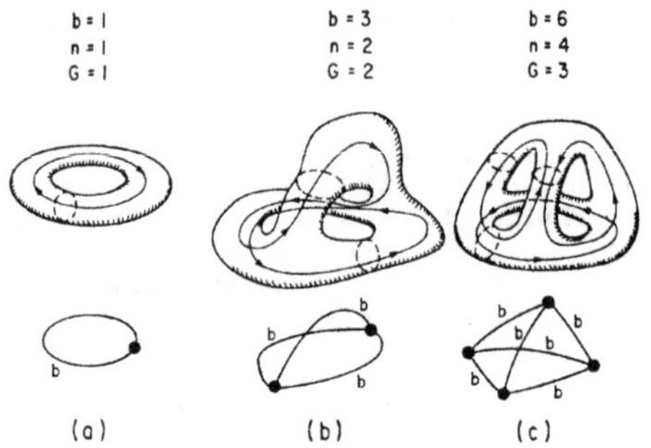

Figure 9. Simple shapes and their branch-node chart used to compute the genus (after Macdonald et al. [48]).

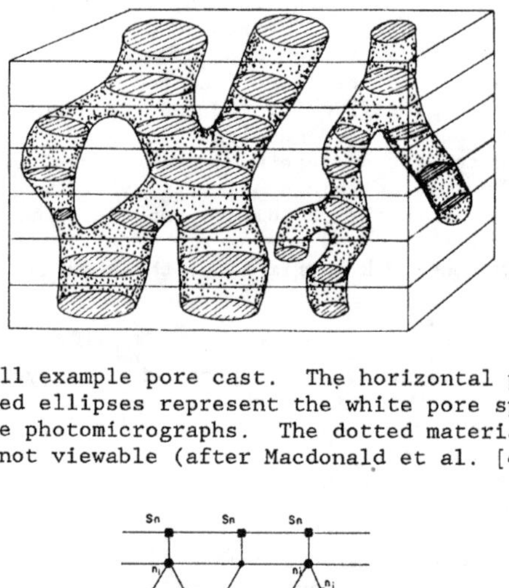

Figure 10(a). A small example pore cast. The horizontal planes are serial sections. The shaded ellipses represent the white pore space features which would be seen in the photomicrographs. The dotted material is pore space between planes and not viewable (after Macdonald et al. [48]).

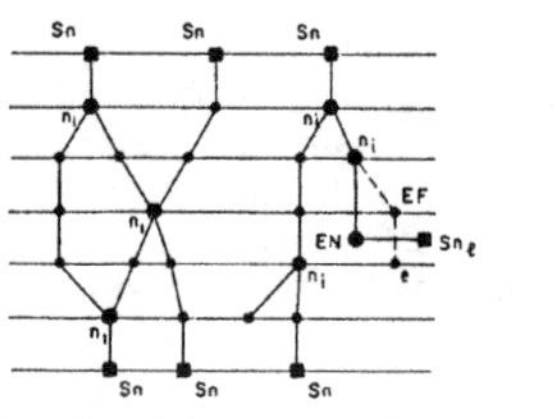

$$G_{MIN} = (b_i + b_\ell) - (n_i + EN + Sn + Sn_\ell) + N =$$
$$= (12+1) - (6+1+6+1) + 2 = 13 - 14 + 2 = 1$$

$$G_{MAX} = (b_i + b_\ell) - (n_i + EN) = (12+1) - (6+1) =$$
$$= 13 - 7 = 6$$

- • PORE SPACE FEATURE
- ● INTERNAL NODE OF TYPE n_i or EN AS INDICATED
- ■ SURFACE NODE OF TYPE Sn or Sn_ℓ

Figure 10(b). The branch-node chart for the pore space in Figure 10(a), showing the types of nodes and boundary features used (after Macdonald et al. [48]).

the lateral surface nodes are not and they must be obtained by means of the overlapping criterion. Much of the study [48] deals with the problem of edge effects in the construction of the branch-node chart which are very important owing to the small size of the sample. Larger sample, however, would result in much increased computer time and decreased resolution of the features. The maximum possible value of the genus G_{max} is obtained by connecting all the surface nodes to one external node. The least possible value of the genus, G_{min}, is calculated by not connecting any of the surface nodes to an external node. In the Berea sample $G_{max} = 593$ and $G_{min} = 420$. Dividing these values by the sample volume of 10.4×10^8 μm^3 the genus per unit volume can be calculated. This yields a genus of about 5×10^{-7} per μm^3, or a genus of 2 per 4×10^6 μm^3 which corresponds to a sphere of a diameter of about 200 μm^3, which is the size of an average grain in the Berea sandstone sample. The pertinent data on the Berea sandstone sample (Berea 2c) and one another smaller, preliminary sample (Berea 1xx) are listed in Table 2. The genus was determined versus the volume of the sample section-by-section with the interesting result that past a certain minimum volume the genus is a linear function of the volume (see Figure 11). The slope of the line gives the best estimate of the genus per unit volume. It is logical that linearity could not exist if the pore topology of the sample had varied in the direction of grinding and polishing, i.e., normal to the planes of sectioning. Although pore geometry, e.g., pore sizes may be different for the same topology, nevertheless in one and the same sample direct proportionality between genus and volume is a strong indication of homogeneous, or uniform, pore structure.

The work published in [48] is an improvement over that of Pathak et al. [44] who performed a manual trace and count of branches and nodes.

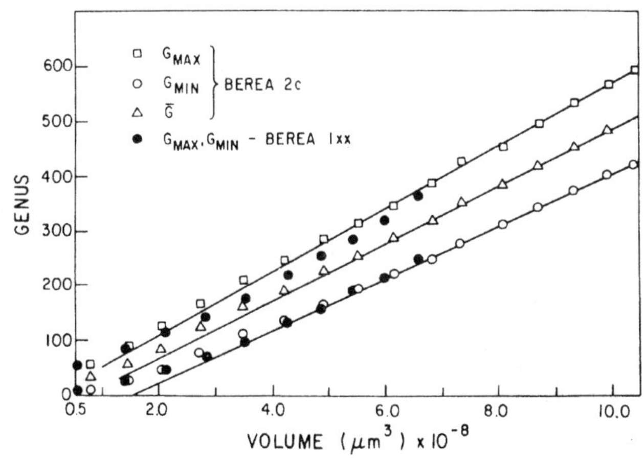

Figure 11. Genus vs. volume of Berea sandstone samples.

3.2. Pore Structure Determination Based on Computer Reconstruction

In a series of articles Lin and coworkers [45,46,53] presented a deterministic approach to modelling the three-dimensional pore and grain geometry and pore network topology, based on computer reconstruction of serial sections. Lin and Perry [46] used a pore (or grain) surface triangulation technique as a shape descriptor, which gives the following

TABLE 2. Berea sandstone samples (after Macdonald et al.[48].

Property	Berea 2c	Berea 1xx
Cross-section	1310 μm x 1040 μm = 1.36 x 10^6 μm^2	1350 μm x 950 μm = 1.28 x 10^6 μm^2
Total depth	762 μm	514 μm
No. of serial sections	78	50
Avg. spacing, $\Delta \bar{z}$	9.9 μm	10.5 μm
Volume	10.4 x 10^8 μm^3	6.6 x 10^{-8} μm^3
No. of features in sample	3564	2583
Avg. grain size	≈ 200 μm	≈ 200 μm

parameters: surface area, Gaussian curvature, genus, and aspect ratio of the pore. The aspect ratio was obtained by using a spheroidal model. In their article, however, they pointed out that their method is not suitable for modelling the pore network. The method used by Lin and Cohen [45] is similar to the one described by DeHoff et al. [42] and Pathak et al. [44]. In another study, Lin [53] carried out three-dimensional measurements in the pore space in the direction of the three orthogonal axes and then used these as parameters for pore models, consisting of ellipsoids, or elliptical cylinders or double elliptical cones.

The same set of 78 photomicrographs, representing serial sections through a Berea sandstone sample, that were processed in [48], were used for locating, at random, points in the digitized three-dimensional pore space and measuring, in three orthogonal directions the lengths of straight lines passing through each point [50] as illustrated in Figure 12.

One important lesson learned from the study reported in [50] is that unrealistically large pore sizes are found by measuring the lengths of straight lines passing through the pore space because the lines pass through more than one pore.

3.3. Method of locating pore throats in computer reconstruction

The approach followed in [54,55] was to first locate the pore throats and then, by symbolically closing them, define the pore bodies. A pore throat is defined as a local minimum in the "size" of pore space which thus separates two pore bodies from one another. As discussed earlier in this

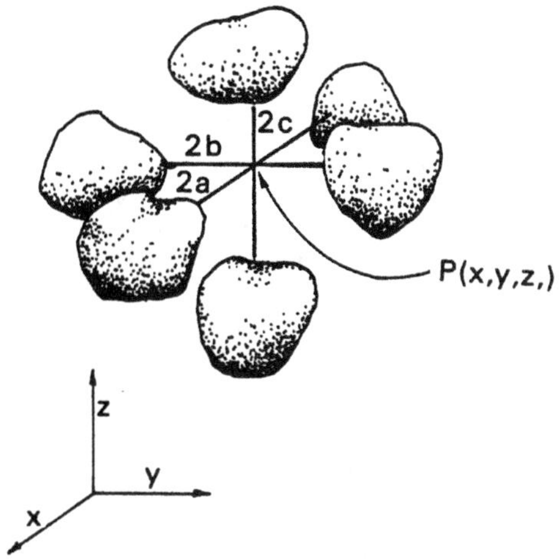

Figure 12. Determination of pore size in three orthogonal directions at a randomly chosen point P(x,y,z) (after Yanuka et al. [50]).

paper, the most practical definition of pore radius is twice the minimum value of the ratio: area of cross section passing through a fixed point in the pore space-to-the perimeter of this section (see Equation (6)). The ratio has, in general, different values for different orientations of the sectioning plane passing through the same fixed point and its minimum value is defined here as the "hydraulic radius" r_H of the pore at that point. Hence, pore throats, by definition, correspond to minima of the hydraulic radius r_H (minima of minima!) In principle, then, a pore throat could be located by passing series of parallel sectioning planes of all possible different orientations through the pore space, calculating the cross section-to-perimeter ratio and keeping track of the location of each section through the pore space. Proceeding along each pore channel, for every fixed orientation of the sectioning plane there will be local minima of the cross section-to-perimeter ratio, indicating the presence of a pore throat. The true throat size, i.e. the hydraulic diameter = four times the hydraulic radius, at a given location is found by varying the orientation of the sectioning plane over all possible angles until the least value of the cross section-to-perimeter ratio is found near that location. The least value is, by definition, the hydraulic radius of that particular throat. All the throats and their hydraulic diameters can be found, rigorously speaking, only by using parallel series of sectioning planes of all possible orientations with a sufficiently small interplanar distance.

The ideal way of locating pore throats, described above, was replaced with the practical way of scanning the computer reconstruction of pore structure with a few sets of parallel planes of distinct, different orientations. For each scan there is a set of potential pore throats. These sets are compared in order to identify the true throats, using the principles outlined earlier. At the time of writing the work of improving this technique is still in progress, because a number of throats appear to have been missed by the scanning and some other throats exhibit anomalous behavior.

After identifying a pore throat, a set of solid matrix voxels have been introduced in its place, thus separating the two adjacent pore bodies. The coordination number, i.e. the number of throats belonging to each pore has been determined. The volume of each pore body was directly obtained by adding up the volumes of the voxels contained in it. In addition, the dimensions of the smallest rectangular parallelepiped completely containing each pore body have also been determined. Some of the results are presented in Figures 13-15.

Figure 13 shows the frequency distribution of the coordination numbers. The average coordination number is 2.9.

The frequency distribution of pore throat diameters is presented in Figure 14. The average neck diameter was found to be 15.5 μm. The throat size distribution found is very close to the corresponding distribution used with good results in the network simulations [14,16,22,23] (see Figure 3).

The frequency distribution of pore bodies, modelled as cubes, is shown in Figure 15. The number average pore body diameter, based on the cube model was found to be about 29 μm. The frequency distribution of pore bodies, modelled as spheres yields an average pore body diameter of about 36 μm. These values are very close to the distribution shown in Figure 3, except for the apparent presence of a relatively small number of very large pores. These may be due to not having located all the pore throats and, therefore, counting several distinct pore bodies as one and the same pore. The relatively small number of very large coordination numbers in Figure 13 is probably due to the same error.

4. MICROSCOPIC DISTRIBUTION OF THE WETTING AND THE NONWETTING PHASE IN IMMISCIBLE DISPLACEMENT

The distribution of the phases in the pore space in immiscible displacement is of great interest. It depends, in addition to the saturation, on the wettability conditions, the history (including the effect of parameters such as the capillary number, the viscosity ratio and the individual viscosities) and last, but not least, the pore structure. Pioneering work in this area has been reported in [56] in Berea sandstone for the special case of strong preferential wettability and quasistatic displacement (vanishingly small capillary number). The technique used consists of "phase immobilization."

A suitable pair of immiscible fluids have been used as the wetting and the nonwetting phase, one of which can be conveniently solidified in situ and the other one can be readily removed from the pore space afterwards. The "empty pore space" permits conventional permeability measurements to be carried out instead of the usual steady state relative permeability measurements. Resistivity index measurements can also be performed if the "empty pore space" is filled with an electrolyte solution. After carrying out these measurements the "empty pore space" is filled with another liquid of the type that can afterwards be solidified in situ. The rock matrix was also replaced with epoxy resin after etching with hydrofluoric acid and finally has either been polished or thin sections have been prepared. The

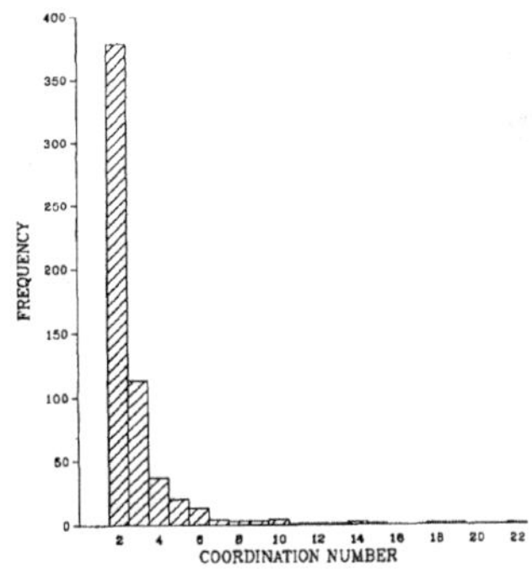

Figure 13. Frequency distribution of coordination numbers (after Kwiecien [54]).

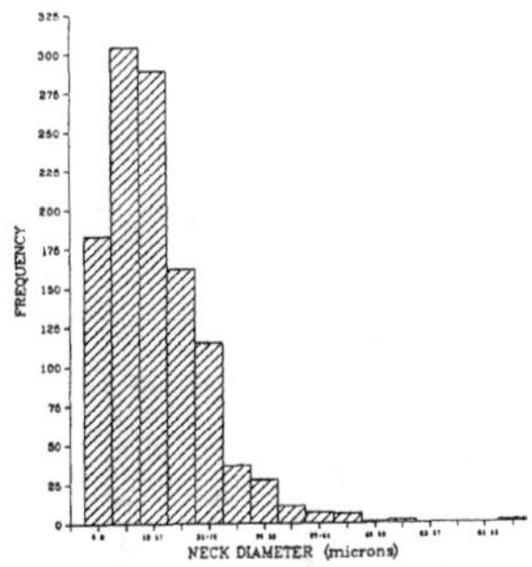

Figure 14. Frequency distribution of neck (pore throat) diameters (after Kwiecien [54]).

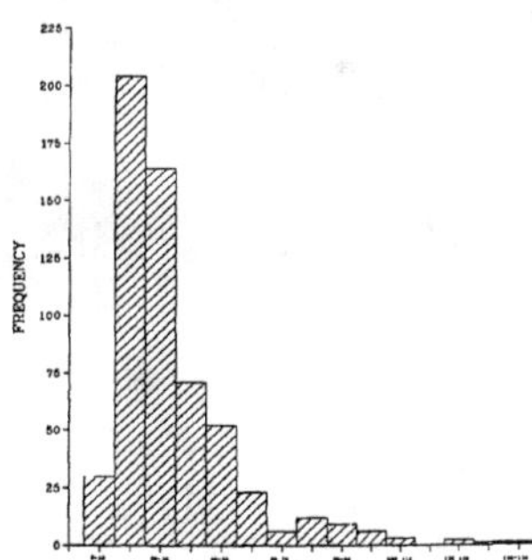

Figure 15. Frequency distribution of pore body diameters. Pore bodies modelled as cubes (after Kwiecien [54]).

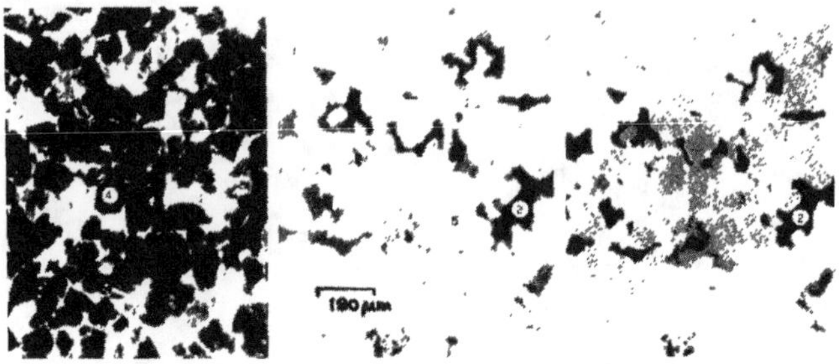

Figure 16. Microscopic distribution of fluids in a typical thin section of Berea sandstone at a wetting phase saturation of 53%, showing (a) the wetting phase only (white portion); (b) the nonwetting phase only (dark portion); (c) the wetting phase, nonwetting phase and rock (white, black and grey portions, respectively) (after Yadav et al. [56]).

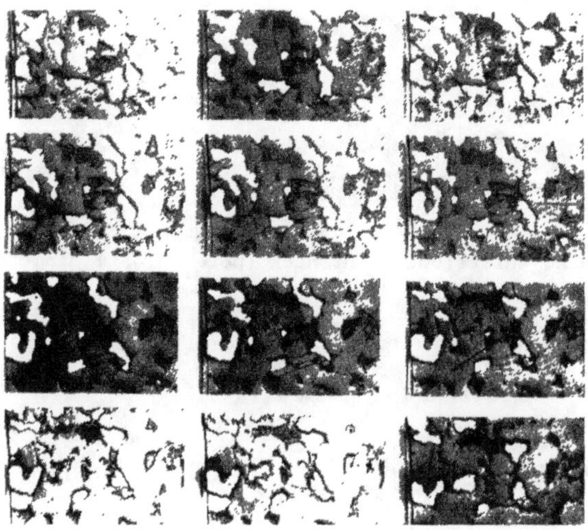

Figure 17. Twelve consecutive serial sections of etched Berea sandstone at about 10 μm apart, seen under normal light. The white portions are Wood's metal, representing the nonwetting phase. The dark grey portions are resin ERL 4206, replacing ethylene glycol, the wetting phase. The lighter grey areas are Buehler resin, replacing the rock that was etched away (after Yadav et al. [56]).

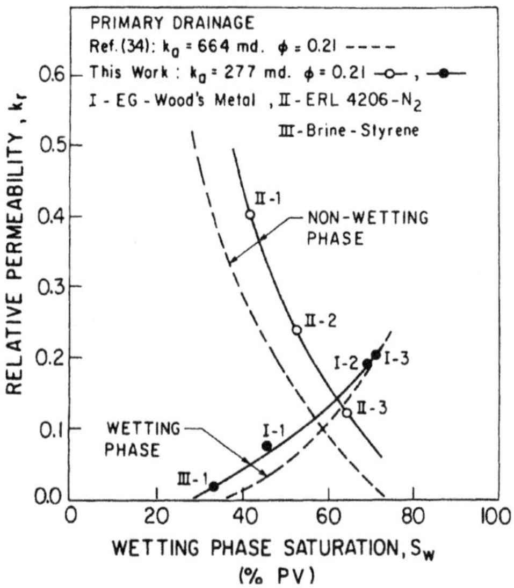

Figure 18. Relative permeability vs. saturation curves for Berea sandstone sample obtained using phase immobilization technique compared with the curves obtained by the usual steady state technique [24] (after Yadav et al. [56]).

following fluid pairs, representing the wetting and the nonwetting phase, respectively, have been used. System I - ethylene glycol/Wood's metal (alloy 158); System II - epoxy resin ERL 4206™/N_2 gas; System III - brine/styrene (containing benzoyl peroxide as the catalyst).

In Figure 16 a thin section shows the microscopic phase distribution in primary drainage obtained with the help of System II. The nonwetting phase channels were impregnated with Resin 301™, containing solvent blue dye. In Figure 16(a) a UV light source, in Figure 16(b) normal light source was used. Figure 16(c) is a superimposition of Figures 16(a) and 16(b), achieved through the controlled use of both light sources.

In Figure 17 twelve consecutive serial sections, prepared by the grinding and polishing procedure, are shown. System I was used in primary drainage. The wetting phase saturation is 58% pore volume. The ethylene glycol was replaced with ERL 4206.

Finally, in Figure 18 "relative permeability curves" obtained by conventional permeability measurements in the presence of another, immobilized phase are compared with the conventional steady state relative permeability curves measured in a similar Berea sandstone [24]. The agreement is very good.

5. DISCUSSION AND CONCLUSIONS

Throughout the present review the position has been taken by the author that any model of pore structure should have for its first and foremost aim to approximate the significant features of the real pore structure of the sample of the porous medium as closely as possible and necessary. Those details of the pore structure which have no or only very little bearing on the transport properties of the medium are to be omitted as they would unnecessarily increase the complexity of the model without any concomitant improvement in its predictive ability. The irrelevancy of certain details may even lead to predictions that are at variance with experience in some cases whereas there may be other cases when a certain peculiar behavior of the medium can be explained only with the help of certain pore structure features that for most other purposes are irrelevant. As has always been the case in mathematical modeling of physical phenomena, judgment must be used in deciding what features to retain in the model and what other features to omit. While admittedly there exists a "grey zone" of uncertainty when deciding where to draw the fine line between what is kept and what is discarded as superfluous, that does not in the least put in jeopardy the requirement that <u>any</u> model of pore structure should account for the main features of the real pore structure which determine the collection of the most important transport properties of the medium. A rough comparison may be made with the blueprint of a building where all the essential constructional features are shown, however, without specifying the location of every hole to be drilled in the walls, the quality of wall surface, etc.

At the present there is no proven model that would be able to simulate all transport properties and, therefore, there is no guarantee that the following requirements regarding a good pore structure model would be sufficient or necessary. In any event, they are most likely to be necessary to do the job:

i. a 3-dimensional network of pore bodies connected by pore throats, representing the main skeleton of pore structure of the medium;
ii. a representative coordination number distribution and the connectivity of the network;
iii. representative pore body and pore throat shapes (aspect ratios);
iv. representative pore body size and pore throat size distributions;
v. a representative correlation (if any present) between the pore throat sizes and the sizes of the two pore bodies connected by a throat;
vi. similar properties of secondary networks of smaller (micro-) pores if such are present (e.g., cementing clays in sandstones or micropores present in the individual particles of aggregates).

Additional requirement for the purpose of predicting surface transport properties is the quality of the pore surface (rugosity).

The only way it appears possible to obtain all this body of information is by visualization of the pores.

This is the reason why the author has chosen the route of 3-dimensional computer reconstruction of the pore structure from serial sections of the sample. This technique, however, has the drawback of requiring lengthy and painstakingly careful sample preparation and it is restricted to pore sizes down to about 5 μm, because the precision of preparing serial sections is a few microns at best in the interplanar distance between consecutive sections. Hence there is no point in trying to prepare serial sections with a spacing of less than about 10 μm. Additional difficulties must be overcome in making the spacings as uniform and the sections as plane and parallel as possible. Microtomography can cut slices 1 μm apart but the accuracy of sectioning is not sufficient for quantitative reconstruction work. NMR imaging is an other possible route, but at the time of writing there are no published pore structure reconstructions based on this approach.

It would be of great advantage to be able to avoid having to make serial sections and work instead with 2-dimensional sections from which the 3-dimensional pore structure would be reconstructed. A major contribution to this technique has been made by Quiblier [61] who has shown that it is possible to perform image analysis on thin sections of porous media, resulting in an autocorrelation function and a probability density function of optical densities in the micrograph analyzed, and use this information to generate a 3-dimensional pore structure that may have the same morphology as the sample from which the thin section was prepared. It appears that this technique would actually be simpler to use in the case of polished sections containing Wood's metal in the pore space because of the higher contrast between pore space and solid rock matrix. The method needs much more extensive testing before it can be accepted and used routinely, but the time and money would be spent on a very worthwhile project because it has a good chance of achieving its objective, i.e., to have a relatively fast and convenient, routine method of determining 3-dimensional pore structures of porous media over a wide range of pore sizes, down to at least 0.1 μm.

6. REFERENCES

[1] Klinkenberg, L.J. (1957) 'Pore size distribution of porous media and displacement experiments with miscible liquids', Pet. Trans. Am. Inst. Min. Eng. 210, 366.

[2] Dullien, F.A.L. (1979) Porous Media: Fluid Transport and Pore Structure, Academic Press, New York.

[3] Carman, P.C. (1941) 'Capillary rise and capillary movement of moisture in fine sands', Soil Sci. 52, 1.

[4] Fatt, I. (1956) 'The network model of porous media I. Capillary pressure characteristics', Pet. Trans. AIME 209, 114.

[5] Fatt, I. (1956) 'The network model of porous media II. Dynamic properties of singe size tube network', Pet. Trans. AIME 207, 160.

[6] Fatt, I. (1956) 'The network model of porous media III. Dynamic properties of networks with tube radius distribution', Pet. Trans. AIME 207, 164.

[7] Simon, R. and Kelsey, F.J. (1971) 'The use of capillary tube networks in reservoir performance studies: I. Equal-viscosity miscible displacements', Soc. Petroleum Engrs. J. 11, 99.

[8] Simon, R. and Kelsey, F.J. (1972) 'The use of capillary tube networks in reservoir performance studies: II. Effect of heterogeneity and mobility on miscible displacement efficiency', Soc. Petroleum Engrs. J. 12, 345.

[9] Broadbent, S.R. and Hammersley (1957) 'Percolation processes. I. Crystals and Mazes', Proc. Cambridge Phil. Soc. 53, 629.

[10] Shante, V.K.S. and Kirkpatrick, S. (1971) 'An introduction to percolation theory', Adv. Phys. 42, 385.

[11] Kirkpatrick, S. (1973) 'Percolation and conduction', Rev. Math. Phys. 45, 574.

[12] Chatzis, I. and Dullien, F.A.L. (1977) 'Modeling pore structure by 2-D and 3-D networks with application to sandstones', J. of Can. Petr. T. 16, 97.

[13] Chatzis, I. and Dullien, F.A.L. (1978) 'A network approach to analyze and model capillary and transport phenomena in porous media', Proceedings of the IAHR Symposium "Scale Effects in Porous Media" held in Thessaloniki, Greece, Aug. 29-Sept. 1.

[14] Chatzis, I. and Dullien, F.A.L. (1985) 'The modelling of mercury porosimetry and the relative permeability of mercury in sandstones using percolation theory', I.C.E. 25, 1, 47.

[15] Wardlaw, N.C. (1976) 'Pore geometry of carbonate rocks as revealed by pore casts and capillary pressure", Am. Ass. Pet. G. Bull 60 (2), 245.

[16] Diaz, C.E., Chatzis, I., and Dullien, F.A.L. (1987) 'Simulation of capillary pressure curves using bond correlated site percolation on a simple cubic network', Transport in Porous Media 2, 215.

[17] Lenormand, R. and Zarcone, C. (1984) 'Role of roughness and edges during imbibition in square capillaries', SPE 13264. Paper presented at the 59th Annual Technical Conference and Exhibition of the Society of Petroleum Engineers of AIME, held in Houston, TX, Sept. 16-19.

[18] Li, Y. and Wardlaw, N.C. (1986) 'Mechanisms of nonwetting phase trapping during imbibition at slow rates', J. Colloid Interface Sci. 109, 473.

[19] Chatzis, I. and Dullien, F.A.L. (1983) 'Dynamic immiscible displacement mechanisms in pore doublets: Theory versus experiment', J. Colloid Interface Sci. 91, 199.

[20] Dullien, F.A.L., Lai, F.S.Y., and Macdonald, I.F. (1986) 'Hydraulic continuity of residual wetting phase in porous media', J. Colloid Interface Sci. 109, 201.

[21] Dullien, F.A.L., Zarcone, C., Macdonald, I.F., Collins, A., and Bochard, R.D.E. (1989) 'The effects of surface roughness on the capillary pressure curves and the heights of capillary rise in glass bead packs', J. Colloid Interface Sci. 127, 362.

[22] Kantzas, A. and Chatzis, I. (1988) 'Network simulation of relative permeability curves using a bond correlated-site percolation model of pore structure', Chem. Eng. Comm. 69, 191.

[23] Kantzas, A. and Chatzis, I. (1988) 'Application of the preconditioned conjugate gradient method in the simulation of relative permeability properties of porous media', Chem. Eng. Comm. 69, 169.

[24] Shankar, P.K. and Dullien, F.A.L. (1981) 'Experimental investigation of two-liquid relative permeability and dye adsorption capacity versus saturation relationships in water-wet and dry-film-treated sandstone samples', in D.O. Shah (ed.), 3rd International Conference on Surface and Colloid Science: Surface Phenomena in Enhanced Oil Recovery, Plenum Press.

[25] Winterfeld, P.H., Scriven, L.E., and Davis, H.T. (1981) 'Percolation and conductivity of random two-dimensional composites', J. Phys. Chem. 14, 2361.

[26] Chandler, R., Koplik, J., Lerman, K., and Willemsen, J.F. (1982) 'Capillary displacement and percolation in porous media', J. Fluid Mec. 119, 249.

[27] Mohanty, K.K. and Salter, S.J. (1982) 'Flow in porous media II. Pore level modeling', SPE paper no. 11018, presented at the 57th Annual Fall Technical Conference and Exhibition of SPE of AIME held in New Orleans, LA (Sept. 26-29).

[28] Wilkinson, D. and Willemsen, J.F. (1983) 'Invasion percolation: A new form of percolation theory', J. Phys. A: Math Gen. 16, 3365.

[29] Heiba, A.A., Sahimi, M., Scriven, L.E., Davis, H.T. (1982) 'Percolation theory of two phase relative permeability', SPE Paper No. 11015, presented at the 57th Annual Fall Technical Conference of SPE-AIME, New Orleans, (Sept. 26-29).

[30] Larson, R.G., Scriven, L.E., and Davis, H.T. (1981) 'Percolation theory of two phase flow in porous media', Chem. Engng. Sci. 36, 75.

[31] Koplik, J., Lin, C., and Vermette, M. (1984) 'Conductivity and permeability from microgeometry', J. Appl. Phys. 56, 3127.

[32] Guyon, E., Hulin, J.P., and Lenormand, R. (1984) 'Application de la percolation à la physique des milieux poreux', Annales des Mines, mai-juin, p. 17.

[33] Wilkinson, D. and Barsony, M. (1984) 'Monte Carlo study of invasion percolation clusters in two and three dimensions', J. Phys. A: Math. Gen. 17, L129.

[34] Wilkinson, D. (1984) 'Percolation model of immiscible displacement in the presence of buoyancy forces', Physical Review A 30, 520.

[35] Wilkinson, D. (1986) 'Percolation effects in immiscible displacement', Physical Review A 34, 1380.

[36] Dias, M.M. and Wilkinson, D. (1986) 'Percolation with trapping', J. Phys. A 19, 3131.

[37] de Arcangelis, L., Koplik, J., Redner, S., and Wilkinson, D. (1986) 'Hydrodynamic dispersion in network models of porous media', Physical Review Letters 57, 996.

[38] Li, Y., Laidlaw, W.G., and Wardlaw, N.C. (1986) 'Sensitivity of drainage and imbibition to pore structures as revealed by computer simulation of displacement process', Advances in Colloid and Interface Sci. 26, 1.

[39] Lenormand, R., Touboul, E., and Zarcone, C. (1988) 'Numerical models and experiments on immiscible displacements in porous media', J. Fluid Mech. 189, 165.

[40] Fischmeister, H.F. (1974) 'Pore structure and properties of materials', in Proc. Int. Symp. RILEM/UPAC, Prague, September 18-21, 1973, Part II, C435, Academic, Prague.

[41] DeHoff, R.T. and Rhines, F.N. (eds.) (1968) Quantitative Microscopy, McGraw-Hill Book Co., N.Y.

[42] DeHoff, R.T., Aigeltinger, E.H., and Craig, K.R. (1972) 'Experimental determination of the topological properties of three-dimensional micro-structures', J. Microsc. 95, 69.

[43] DeHoff, R.T. (1983) 'Quantitative serial sectioning analysis: Preview', J. Microsc. 131, 259.

[44] Pathak, P., Davis, H.T., and Scriven, L.E. (1982) 'Dependence of residual nonwetting liquid on pore topology', SPE Preprint 11016, 57th Annual SPE Conf., New Orleans.

[45] Lin, C. and Cohen, M.H. (1982) 'Quantitative methods for microgeometric modeling', J. Appl. Phys. 53, 4152.

[46] Lin, C. and Perry, M.J. (1982) 'Shape description using surface triangularization', Proc. IEEE Workshop on Computer Visualization, N.H.
[47] Kaufmann, P.M., Dullien, F.A.L., Macdonald, I.F., and Simpson, C.S. (1983) 'Reconstruction, visualization and topological analysis of sandstone pore structure', Acta Stereol. 2 (Suppl. I), 145.
[48] Macdonald, I.F., Kaufmann, P., and Dullien, F.A.L. (1986) 'Quantitative image analysis of finite porous media. I. Development of genus and pore map software' J. Microsc. 144, 277; 'II. Specific genus of cubic lattice models and Berea sandstone', ibid. 144, 297.
[49] Barrett, L.K. and Yust, C.S. (1970) 'Some fundamental ideas in topology and their application to problems in metallography', Metallography 3, 1.
[50] Yanuka, M., Dullien, F.A.L., and Elrick, D.E. (1986) 'Percolation processes and porous media. I. Geometrical and topological model of porous media using a three-dimensional joint pore size distribution', J. Colloid Interface Sci. 112, 24.
[51] Ridgway, H.L. and Tarbuk, K.J. (1967) 'The random packing of spheres', Brit. Chem. Eng. 12, 384.
[52] Kruyer, S. (1958) 'The penetration of mercury and capillary condensation in packed spheres', Trans. Faraday Soc. 54, 1758.
[53] Lin, C. (1983) 'Shape and texture from serial contours', J. Intl. Assoc. Math. Geol. 15, 617.
[54] Kwiecien, M.J. (1987) 'Determination of pore size distributions of Berea sandstone through three-dimensional reconstruction', M.A.Sc. Thesis, University of Waterloo.
[55] Kwiecien, M.J., Macdonald, I.F., and Dullien, F.A.L. 'Three-dimensional reconstruction of porous media from serial section data', submitted to J. Microsc.
[56] Yadav, G.D., Dullien, F.A.L., Chatzis, I., and Macdonald, I.F. (1987) 'Microscopic distribution of wetting and nonwetting phases during immiscible displacement', SPE Reservoir Engng. 2, 137.
[57] Yuan, H.H. and Swanson, B.F. (1989) 'Resolving pore-space characteristics by rate-controlled porosimetry', SPE Formation Evaluation 4, 17.
[58] Dullien, F.A.L. (1975) 'New permeability model of porous media', AIChE J. 21, 299.
[59] van Brakel, J. (1975) 'Pore space models for transport phenomena in porous media. Review and evaluation with special emphasis on capillary liquid transport', Powder Techn. 11, 205.
[60] van Brakel, J. and Heertjes, P.M. (1977) 'Capillary rise in porous media. Part I: A problem', Powder Techn. 16; 'Part II: Secondary phenomena', Ibid. 16, 83; 'Part III: Role of the contact angle', Ibid. 16, 91.
[61] Quiblier, J.A. (1984) 'A new three-dimensional modeling technique for studying porous media', J. Colloid Interface Sci. 98, 84.
[62] Dullien, F.A.L. (1988) 'Two-phase flow in porous media', Chem. Eng. Technol. 11, 407.

Transport in Porous Media 6: 607–626, 1991

© 1991 *Kluwer Academic Publishers. Printed in the Netherlands.*

SCALING MIXING DURING MISCIBLE DISPLACEMENT IN HETEROGENEOUS POROUS MEDIA

Robert A. Greenkorn

John S. Haselow

School of Chemical Engineering
Purdue University
West Lafayette, IN 47907

Abstract

This paper discusses scaling of mixing during miscible flow in heterogeneous porous media. In large field systems dispersivity appears to depend on system length due to heterogeneities. Three types of scaling are discussed to investigate the heterogeneous effects. Dimensional analysis of mixing during flow through geometerically scaled heterogeneous models is illustrated using measured dispersion. Fractal analysis of mixing in statistically scaled heterogeneous porous media is discussed. Analog scaling of pressure transients in heterogeneous porous media is suggested as an in-situ method of estimating dispersion.

Key Words: dispersion, dispersivity, heterogenity, miscible, porous media, scaling

* Currently Environmental Sciences Division, Savannah River Laboratory, Aiken, SC 24808

0. Notation

L=Length; M=mass; t=time, (1) indicates dimensionless

a	dispersivity (L)	V	local velocity (L/t)
c	concentration (1).	v	velocity (L/t)
c_1	fluid compressibility (Lt^2/M)	$\bar{v}$	time averaged velocity (L/t)
D	dispersion (L^2/t)	W	width (L)
D	fractional dimension (1)	x	coordinate (L)
d	Euclidean dimension (1)	Y	$Y=\ln \bar{k}$ (1)
$\bar{d}$	average particle size (L)	y	coordinate (L)
g	acceleration due to gravity (L/t^2)	ε_c	fractal cutoff (L)
$\bar{k}$	average permeability (L^2)	μ	viscosity (LM/t)
L	length (L)	ϕ	porosity (1)
l_y	correlation scale (1/L)	ρ	density (M/L^3)
N	Number of sites (1)	σ^2	variance (dimension depends on variable)
p	pressure (M/t^2L)	σ	spectral exponent (1)
[R]	random number (1)		
r	radius (L)		
t	time (t)		

1. Introduction

This paper discusses scaling of mixing during miscible flow in heterogeneous porous media with an aim of understanding effects of scale and permeability change on mixing. In the past few years there has been considerable discussion of the applicability of the traditional advection-diffusion equation (Ficks law) for modeling mixing during miscible transport in porous media. It is apparent that mixing during flow in heterogeneous field systems is much larger than might be expected from measurements in homogeneous laboratory models. Further, it is also apparent that under certain conditions the usual advection-diffusion equation does not describe the dominant mixing mechanism. Thus non-Fickian behavior manifests itself as length dependence of the dispersivity. Further, it appears for flow in heterogeneous media that the time (length) to reach quasi-steady state depends on the size of the heterogeneities relative to the system size.

Taylor (1953) described dispersion as a competitive phenomenon between convective and diffusive fluxes. For large $\frac{L}{\bar{v}}$ where L is system length and $\bar{v}$ is average velocity, the diffusive flux is the dominant mechanism. de Josselin de Jong (1956) showed that dispersion is described by a tensor. Nikolaevskii (1962) obtained the advection-diffusion equation and a relation for the dispersion tensor by analgy to diffusion in a statistical model for homogeneous turbulence. Biggar and Nielsen (1962) presented field data that showed mixing in a field to be much larger than expected - due to diffusion (very slow flow). Harleman *et al.* (1963) related dispersion and pore structure. Greenkorn *et al.* (1965) presented data for controlled field and laboratory model experiments that showed agreement if permeability variation was scaled geometrically. Greenkorn and Kessler (1969) report results of Pleshek (1968) and Nieman (1969) that showed that breakout curves in heterogeneous laboratory models are not explained by the advection-diffusion equation. Freeze (1975) suggested a stochastic approach to relate field scale heterogeneity and dispersion. Cushman (1984) introduced the concept that the scale of the system and measurement are fundamentally connected.

There are many approaches to modeling mixing during miscible transport in heterogeneous porous media. The problem of mixing may be studied: by volume averaging the equations of change involving a closure problem for each structure (Plumb and Whitaker 1988); stochastically (Smith and Schwartz, 1980); statistically (Chin, 1986) or deterministically (Cala and Greenkorn, 1986). Selection and use of mathematical models for calculating mixing in heterogeneous systems requires information concerning the relation between mixing mechanism, flow regimes and description of the porous media in question. Data from the scaled laboratory and the field experiments can provide this information.

One of the difficulties in discussing and calculating flow in heterogeneous porous media is the contrasts that result when considering homogeneous and heterogeneous media. Mixing in heterogeneous media is larger than one would predict on the basis of homogeneous results. There is some confounding since as the scale of observation changes we see different mechanisms of mixing and the mechanisms may have a homogeneous and heterogeneous nature. There are regimes that affect mixing due to different mechanisms at each scale. At the molecular level the mixing results from molecular diffusion. It is possible to have heterogeneous effects at this level due to local variation in the number density of the molecules. At the microscopic or pore size level mixing is due to interparticle variation in convective flux. Variation of the surface of the pores can create a heterogeneous effect. At the macroscopic level mixing is due to gradual gradient changes in pore sizes. At this level there may be a variation in permeability from point-to-point but a single average permeability (single distribution) will represent the conductance of the medium. Two media of the same average permeability but with different grade (variance of the permeability distribution) will give different mixing. Thus if the two different media are side-by-side in a flow situation, the media will be heterogeneous. At the megascopic level the mechanism is convective flux due to significant (usually abrupt) change in average permeability of the media. For example, gravel surrounded by sand will create streamline changes as fluid passes from sand to gravel.

The megascopic scale is often associated with the field scale-infering large distances (relative to pore sizes) are involved. If the dominant mechanism for mixing is convective flux due to abrupt change of permeability as long as there is no regime change this mixing may be described by dimensionally similar models of various sizes. Thus a laboratory model of 1 square meter with a permeability change may have the same relative mixing effect as a reservoir 10 square kilometers with the same permeability change, scaled geometrically from 1 square meter to 10 square kilometers. If the flow path is large compared to the size of the permeability change, dispersivity is asymptotically constant and the traditional advection-diffusion equation applies. If the flow path is small compared to the size of permeability change, dispersivity appears to be a function of length or time and grows in a power law fashion,

$$a = c\,L^n \tag{1}$$

where a is dispersivity and L is system length. It may be the growth of dispersivity is a succession of quasi-steady states of the advection-diffusion equation and does not necessarily represent non-Fickian behavior.

Self similar features of porous media such as fracture trees or "telescoping" heterogeneities have a fractal nature. If a telescoping view of space creates a different volume occupied the object has fractional dimension - it is fractal. The number of sites, N, as a function of distance, r is

$$N(r) = (\text{constant})\, r^{D-1} \tag{2}$$

where D is the fractional dimension. The constant is characteristic of the fractal object. If the system is not fractal D is equivalent to the Euclidean dimension. If a porous medium behaves as a fractal, there are two approaches that have been suggested to represent heterogeneity. Hewett [1986] describes field measurements of porosity and permeability such that they retain their fractal structure - that is their point variance. Hewett [1986] suggests interpolating well values of porosity or permeability then building the vertical variation back into the values using fractal description from a well log. Calculations of mixing are carried out with a stream tube model. Chang and Yortsos (1988) describe a fractured reservoir as a fractal object embedded into Euclidean matrix and reformulated the equations (Darcy's law and continuity) in terms of fractional dimensions.

2. Scaling

In the following, three types of scaling are discussed. Dimensional analysis of mixing of geometrically scaled heterogeneous porous media models is illustrated using measured dispersion. Fractal analysis of mixing in statistically scaled heterogeneous porous media is discussed and possible experimental measurements described. Analog scaling of pressure transients in heterogeneous porous media is suggested as an in-situ method of estimating dispersion.

At the megascopic scale mixing is due to gross streamline change as bulk fluid moves through a heterogeneity of different permeability, size and shape than the surrounding media. If the regime of fluid flow does not change - that is- the controlling forces in the fluid flow such as the bulk motion do not change when system size changes one observes the same mechanism at different sizes. Thus, it is possible to observe a megascopic scale phenomena in a generalized scaled model much smaller than the original size. In dimensionless modeling of flow in heterogeneous systems the mixing phenomena caused by a heterogeneity can be scaled, dimensionlessly using a small scaled model as long as the same mixing regime, dominates in the model and its prototype.

2.1 Geometric scaling

Geometric scaling is based on the dimensional analysis hypothesis that the mathematical model for a given process is a dimensionally homogeneous equation (Buckingham, 1914). If geometrically similar systems are compared at the same values of the dimensionless groups then

$$\pi_1' = \pi_1 \qquad (3)$$

$$\text{if} \quad \pi_2' = \pi_2, \; \pi_3' = \pi_3, \ldots = \pi_{n-r}$$

applies to similar systems (fractal or Euclidean) and to systems containing heterogeneities. This is the basis for constructing scaled models (models of different sizes) to study physical processes experimentally by scaling down (reservoir to laboratory). It is important in scaling down (or up) that the regime of the process studied does not change. For example, if one is studying mixing in a heterogeneous system dominated by "large-scale" convection the "large-scale" convection must dominate in the smaller laboratory model.

Table 1 - Homogeneous Dispersion Data

Bead Size (μ)	D	a	$\frac{D}{av^{1.2}}$
208-250	$.40v^{1.2}$	.40	2.5
297-415	$.50v^{1.2}$	.50	2.5
590-810	$.55v^{1.2}$	.55	2.6

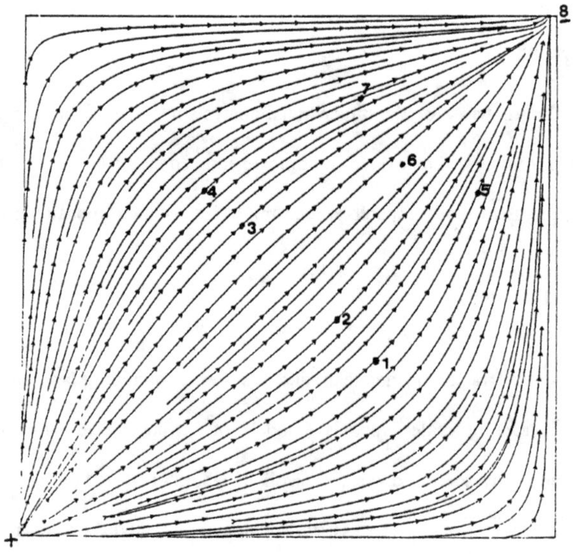

Figure 1 Schematic of Homogeneous Model Showing Calculated Streamlines and Electrode Placement.

The functional form of the dimensionally homogeneous equation describing dispersion is

$$f(L, \bar{d}, \bar{v}, V, \rho, \mu, p, \bar{k}, g, l_y, \sigma_y^2, D, t, c, \phi) = 0 \tag{4}$$

where L is system length, $\bar{d}$ is the average particle size, $\bar{v}$ is the average velocity, V is the local velocity, ρ and μ are fluid density and viscosity, p is pressure, $\bar{k}$ is the average permeability, g is the gravitational constant, l_y is the correlection scale of the logarithm of the permeability ($y = \ln \bar{k}$), σ_y^2 is the variance of y, D is the dispersion coefficient, t is time, c is dimensionless concentration and ϕ is porosity.

The dimensionless functional form of the variables in Eq. (4) is

$$F\left[\frac{t\bar{v}}{L}, \frac{\bar{v}}{V}, c, \phi, \frac{\bar{k}p}{\mu V\bar{d}}, \frac{\bar{k}\rho g}{\mu V}, \frac{V\bar{d}\rho}{\mu}, \frac{D}{\bar{v}\bar{d}}, \frac{\bar{d}}{L}, \frac{\bar{k}}{\bar{d}^2}, \frac{L}{l_y\sigma_y^2}\right] = 0 \tag{5}$$

If "large scale" convective mixing dominates only the first six dimensionless groups (independent variables time and geometry; dependent variables concentration and pressure; porosity, gravity) and the last dimensionless group in Eq. (5) needs to be scaled. The last dimensionless group is the ratio of system size to correlation length scale of the heterogeneity.

2.1.1 Mixing in Laboratory Models

To demonstrate that "large scale" convective mixing can be scaled in laboratory size models we studied mixing in two-dimensional 1/4 5 spot models (radial systems) to answer several questions. Will mixing scale? That is, if we build heterogeneous models of different sizes but the same relative heterogeneity and velocity, will mixing be the same? Will non-Fickian behavior occur? Will it be a funcion of arrangement of permeability?

Figure 1 is a schematic for a homogeneous model showing calculated streamlines and electrode placement for measuring concentration change. The electrodes are placed so, for a homogeneous model, they record similar arrival times. The dispersion can be checked for pairs of electrodes and as a function of dispersion along a streamline. The longitudinal dispersion is calculated by (Bear, 1972)

$$D = \frac{L\sigma^2 V}{2} \tag{6}$$

where the normalized spread of the concentration is

$$\sigma = \frac{t_{84.3} - t_{15.4}}{t_{50}} \tag{7}$$

The time, t_x, is the time since tracer is introduced into the system to reach concentration, x. Three model sizes approximately 10 cm x 10 cm, 20 cm x 20 cm and 40 cm x 40 cm were used. The fluid velocities were scaled such that they are the same in the different size models for all relative positions. The results for the three homogeneous models are given in Table 1. The dispersion for all electrodes is represented by

$$D = a\, v^{1.2} \tag{8}$$

The value of a (dispersivity) was constant throughout the homogeneous models. The dispersivity changes slightly with particle size. Further for each model $\frac{D}{av^{1.2}}$ is a constant which mean the results scale (relative) within experimental error. The heterogeneous results were simulated by adding the convective mixing (due to heterogeneity) at a point to the molecular mixing.

$$x'(t) = w(t) + v_x \Delta t \tag{9}$$

$$y'(t) = y(t) = v_y \Delta t \tag{10}$$

where v_x and v_y were determined from Poisson's equation using a Galerkin finite element solution without inertia. The homogeneous part is

$$x(t+\Delta t) = x'(t) + (2\Delta t\, D_L^{1/2)}\ (1 - ([R]_0^1) \tag{11}$$

$$y(t+\Delta t) = y'(t) + (2\Delta t\, D_T^{1/2})(1 - [R]_0^1) \tag{12}$$

The homogeneous experimental and simulated results are shown in Fig. 2 for all three beadsizes for the 20 cm x 20 cm model for electrode 8. The results are similar for the 10 cm x 10 cm and 40 cm x 40 cm models.

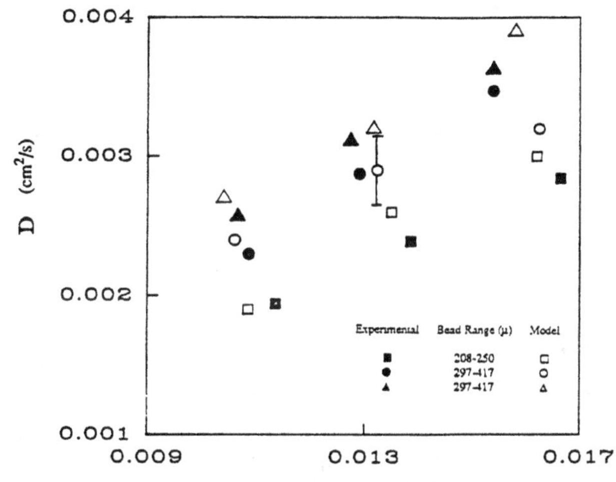

Figure 2 Experimental and Simulated Results for Homogeneous Models at Electrode 8.

The two dimensional heterogeneous models were packed with glass beads in each 1/4 of the model as indicated in the diagram of Fig. 3. The number between the schematic indicates which model was used and the number below indicates the model in which a duplicate was run.

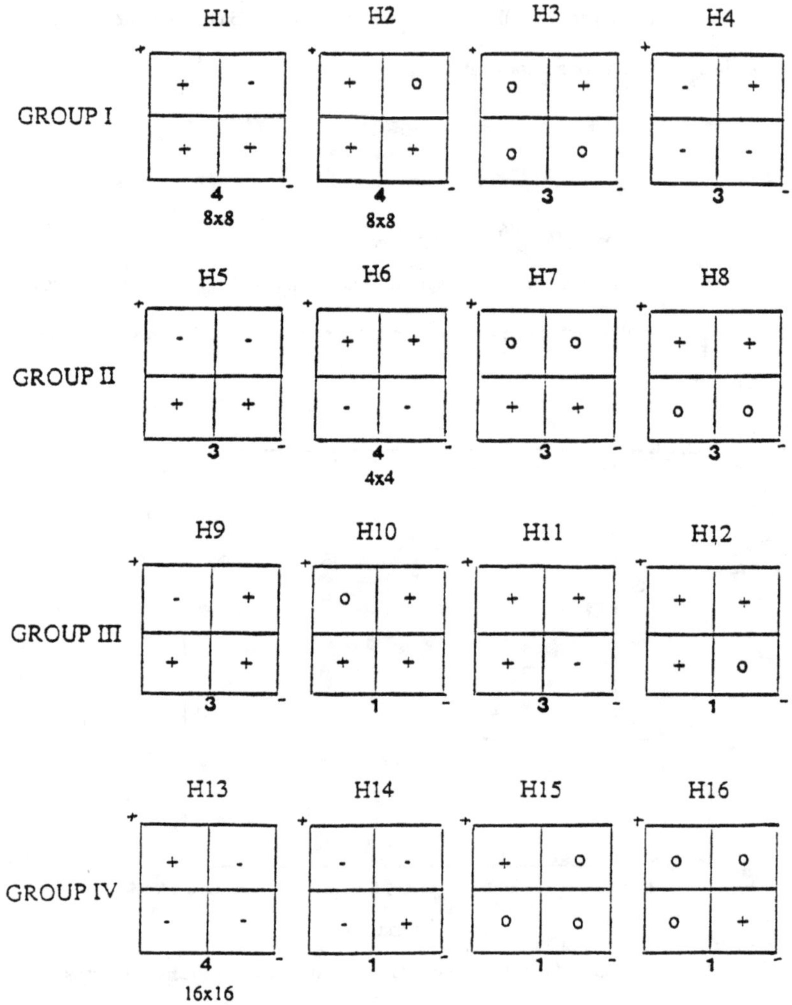

Figure 3 Schematic for all Models

For example, for H1 three model sizes were run and a duplicate was run for the 20 cm x 20 cm; H6 was run on all three sizes with a duplicate in the 10 cm x 10 cm model. If only one model size was run it was in the 20 cm x 20 cm model. Figure 4 is an example of the streamlines for system H1.

A detailed discussion of the experiments is given in Haselow (1988). In summary mixing in the heterogeneous models in Groups I and II is larger than the homogeneous model except in the low flow regime (diffusion controls). The dispersivity fluctuates in the models due to streamline changes The dispersivity is greater past heterogeneous inclusions and varies over the models. For some arrangements of permeability (H1 and H2) mixing is non-Gaussian (Fig. 5). For Groups III and IV mixing is comparable to the homogeneous results but reciprocal arrangements (H9 and H11) yield different mixing. Changing the position of the heterogeneity relative to sinks and sources changes the mixing. The results show for these sizes of models

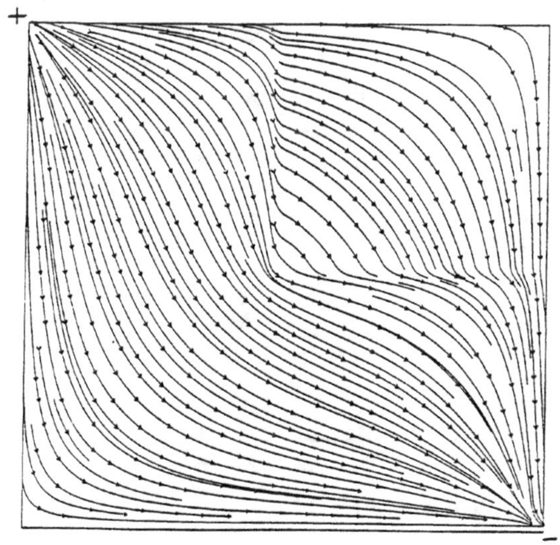

Figure 4 Streamlines for System H1.

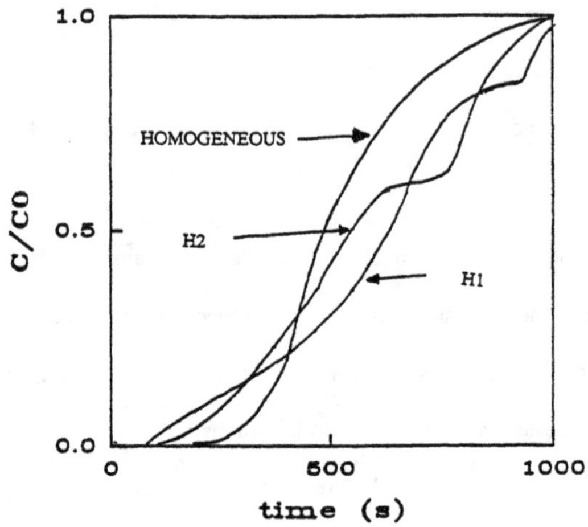

Figure 5 Outlet Concentration Distributions.

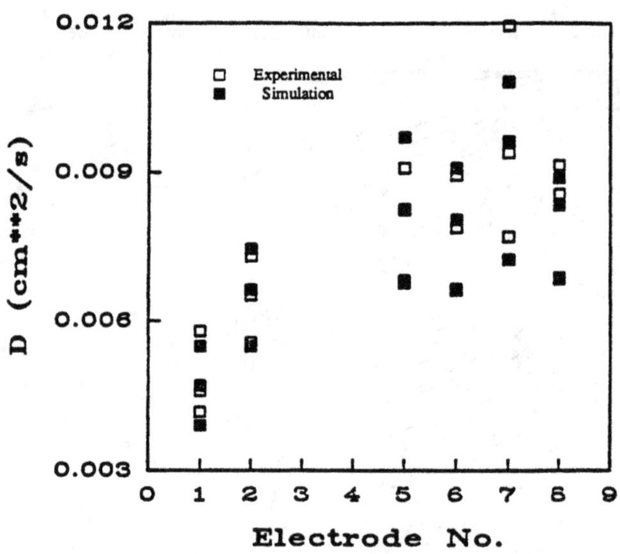

Figure 6 Simulated and Experimental Results for System H1.

that mixing scales, that is, in similar but different size models mixing is relatively the same. In view of the varied results some questions are raised. Is there a perpetual transient state? How can we deal with a system that may change continually between Fickian and non-Fickian?

The simulation of the results for the heterogeneous systems are the same as the homogeneous and give reasonable agreement except for the low flow region (diffusion dominated). Figure 6 is a comparison of the results, except for electrodes 3 and 4 for experiment H1.

The results in these heterogeneous models show mixing does not depend on system size if permeability ratio and geometry of the heterogeneities are similar. Different arrangements of heterogeneities give widely varying mixing and with some arrangements mixing is similar to the homogeneous results. The simulations match except for those locations where diffusion is important.

2.2 Statistical Scaling

Modeling mixing in statistically similar heterogeneous porous media can be approached by modeling the porous media as a fractal object. West and Goldberger (1987) give an example of infinitely nesting spheres to explain the utility of fractional dimensions. The number density of sites in a sphere is represented as

$$N(r) \sim r^D \tag{13}$$

The mass of the sphere if it is filled with sand is proportional to r^3 or

$$M(r) = r^D \tag{14}$$

where D = d = 3. If closer examination of the sphere finds the sand is contained in smaller spheres of radius r/b these smaller spheres will each have a mass 1/a smaller than the total mass.

If examining closer we find the sand is contained in smaller spheres of radius $\frac{r}{b^2}$ then these smaller spheres will each have a mass $1/a^2$ smaller than the total. If we consider the situation for n of these telescoping operations

$$M(r) = M\left[\frac{r}{b^n}\right]a^n = r^D\left[\frac{a^n}{b^{nD}}\right] \tag{15}$$

This relation yields a finite value for total mass as $n \to \infty$ only if

$$D = \frac{\ln a}{\ln b} \tag{16}$$

This fractional value of D is the fractal dimension of the mass distributed through a Euclidean volume of size r. D may be different than d (the Euclidean dimension).

Wheatcraft and Tyler (1988) use fractal scaling to relate distance along a straight line, x_s, to the distance along a fractal path x_f. They assume the fractal paths are statistically similar

$$x_f = \varepsilon_c^{1-D}\, x_s^D \tag{17}$$

where ε_c is the fractal cutoff. The scaling between the straight line and fractal velocity by differentiation is

$$V_f = \varepsilon_c^{1-D}\, D\, x_s^{D-1}\, V_s \tag{18}$$

The dispersivity is defined by the variance of the travel distance

$$\sigma_{xf}^2 = 2ax_s \tag{19}$$

Assuming the porous medium is modeled by a set of fractal streamlines with the variance of the distance traveled given by Eq. (19) then

$$\sigma_{xf}^2 = \sigma_{\varepsilon_c}^{2-2D}\, x_s^{2D} \tag{20}$$

and

$$a = 1/2\, \sigma_{\varepsilon_c}^{2-2D}\, x_s^{2D-1} \tag{21}$$

It should be possible to test the scaling of Eq. (21) by constructing a series of linear models containing porous media of varying permeability (each different) and then measuring the variance of concentration along the models as they are succesively connected together in series. Such a set of experiments is currently being constructed.

2.3 Analog Scaling

Up to this point the discussion has been for homologous scaling - that is, considering different size replicas of the same material. There is also the possibility of analog scaling - study of a physical process which can be modeled with the same equation. Analog scaling of fluid flow using electrical circuits is an example of this type of scaling.

The diffusivity equation has been derived in general for a fractal system by O'Shaughnessy and Procaccia (1985)

$$\frac{\partial \hat{P}(r,t)}{\partial t} = \frac{1}{R^{D-1}} \frac{\partial}{\partial r} \left\{ K\, r^{D-1-\theta} \frac{\partial \hat{P}(r,t)}{\partial r} \right\} \tag{22}$$

where

$$\tilde{d} \equiv \frac{2D}{2+\theta} \rightarrow d \quad \text{if } \theta = 0,\ D = 1, 2, 3 \tag{23}$$

P(r,t) is the probability per site at time t in dr, $\hat{P}$, (r,t) is the envelope of P (r,t), D is the fractal dimension, d is the Euclidean dimension and θ is a spectral exponent If $\hat{P} \equiv p$ (pressure) then K is α $\left[\alpha = \frac{k}{\phi \mu c_1}\right]$ for the fractal object varying with r. If $\hat{P} \equiv c$ (concentration) the K = D (r) - the dispersion of a fractal object varying with r. For simplicity let D = 1, 2, or 3 and $\theta = 0$ so D = d. For simplicity let d = 1 then

$$\frac{p - p_0}{p_1 - p_0} = 1 - \text{erf}\,\frac{x_p}{2\sqrt{\alpha t_p}} \tag{24}$$

For concentration including convection and changing coordinates so $x = x' - v_x t$

$$\frac{c - c_0}{c_1 - c_0} = 1 - \text{erf}\frac{x_c' - v_x t_c}{2\sqrt{D t_c}} \tag{25}$$

If $\dfrac{p - p_0}{p_1 - p_0} = \dfrac{c - c_0}{c_1 - c_0}$ then the pressure change is the analog of the concentration change so

$$\frac{D}{\alpha} = \frac{(x_c' - v_x t_c)^2}{x_p^2}\,\frac{t_p}{t_c} \tag{26}$$

In principle for a porous medium, fractal or otherwise, it may be possible to estimate mixing for a given measured element between two wells, from the transient pressure response, between the wells. That is for a given field system the pressure transient may be used to estimate the dispersion for that same system. It should be possible to test this analog in a linear heterogeneous laboratory model and as well as in a field system.

3. Summary

Three types of scaling - geometric, statistical and analog were discussed for constructing experiments to measure dispersion in heterogeneous porous media systyems. Data for geometrically similar models showed mixing in heterogeneous models can be scaled and non-Fickian behavior occurs in these models. An experimental method is discussed to investigate the fractal nature of heterogeneous systems. Finally analog scaling of pressure transients is suggested as an in-situ method of estimating dispersion.

References

Bear, J., 1972, Dynamics of Fluids in Porous Media, Elsevier, New York.

Biggar, J.W. and D.R. Nielsen, 1962, Some Comments on Molecular Diffusion and Hydrodynamic Dispersion in Porous Media, J. Geophys. Res. 27, 3636-3637.

Buckingham, E., 1914, On Physically Similar Systems; Illustrations of the Use of Dimensional Equations., Phys. Rev. *IV*(4), 345-.

Cala, M.A. and R.A. Greenkorn, 1986, Velocity Effects on Dispersion in Porous Media With a Single Heterogeneity, Water Resour. Res. 22(6) 919-926.

Chang, J. and Y.C. Yortsos, 1988, Pressure Transient Analysis of Fractal Reservoirs, SPE 18170, 63rd Annual Technical Conference, Soc, Pet. Eng., Houston, TX.

Chin, D.A., 1986, Estimation of Dispersion Coefficients in Porous Media, M. Hydraul. Eng.*112*, 591-609.

Cushman, J.H., 1984, On Unifying Concepts of Scale, Instrumentation and Stochastics in the Development of Multiphase Transport Theory, Water Resour. Res.*20*(11), 1668-1672.

de Josselin de Jong, G., 1958, Longitudinal and Transverse Dispersion in Granular Deposits, Trans Amer. Geophys, Union *59*, 67-74.

Freeze, R.A., 1975, A Stochastic Conceptual Analysis of One-Dimensional Groundwater Flow in a Nonuniform Homogeneous Medium, Water Resour, Res. *11*, 725-741.

Gelhar, L.W. and C.L. Amess, Three Dimensional Stochastic Analysis of Macrodispersion in Aquifers, Water. Resour. Res.*19*, 161-180.

Greenkorn, R.A., C.R. Johnson, and R.E. Haring, 1965, Miscible Displacement in a Controlled Natural System, Pet. Trans. AIME, R*234*, 1329-1335.

Greenkorn, R.A. and D.P. Kessler, 1969, Dispersion in Heterogeneous Nonuniform Anisotropic Porous Media, Ind. Eng. Chem. *61*(9), 14-32.

Harleman, D.R.F., P.F. Melhorn, and R.R. Rumer, Jr., 1963, Dispersion-Permeability Correlation in Porous Media, Proc. ASCE J. Hydr. Div. *67*, 67-85.

Haselow, J.S., 1988, Scaling Dispersion during Miscible Displacement in Heterogeneous Porous Media, Ph.D. Thesis, Purdue University, West Lafayette, In.

Hewett, T.A., 1986, Fractal Distributions of Reservoir Heterogeneity and Their Influence on Fluid Transport, SPE 15386 61st Annual Technical Conferences Soc. Pet. Eng., New Orleans, LA.

Nieman, E.H., 1969, Dispersion During Flow in Non-Uniform Heterogeneous Porous Media, M.S. Thesis Purdue University, West Lafayette, In.

Nikolaevskii, V.N., 1959, Convective Diffusion in Porous Media, Prikl. Math. Mech. *23*(6), 1042-1050.

O'Shaughnessy, B.and I. Proaccia, 1985, Diffusion on Fractals, Phys. kRev. A., *32*(5) 3073-3083.

Pleshek, R.C., 1968 dispersion During Flow in Linear Heterogeneous Porous Media, M.S. Thesis, Purdue University, West Lafayette, IN.

Plumb, O.A. and S. Whitaker, 1988, Dispersion in Heterogeneous Porous Media I. Local Volume Averaging and Large-Scale Averaging, Water Resour. Res. 24(7), 913,926.

Smith, L. and F.W. Schwartz, 1980, I. A. Stochastic Analysis of Macrodispersion, Water Resour. Res. *16*(2), 303-313.

Taylor, G. 1953, Dispersion of Soluble Matter in Solvent Flowing Slowly Through a Tube, Proc. roy. Soc. London *A219*, 186-203.

West, B.J. and A.L. Goldberger, 1987, Physiology in Fractal dimensions, Amer. Scientist, *75*, 354-365.

Wheatcraft, S.W. and S.W. Tyler, 1988, An Explanation of Scale-Dependent Dispersivity in Heterogeneous Aquifers Using Concepts of Fractal Geometry, Water Resour. Res. *24*(4), 566-578.

Transport in Porous Media **6**: 627–649, 1991
© 1991 *Kluwer Academic Publishers. Printed in the Netherlands.*

FIXED DOMAIN METHODS FOR FREE AND MOVING BOUNDARY FLOWS IN POROUS MEDIA

J.C. Bruch, Jr.
Department of Mechanical and Environmental Engineering
University of California, Santa Barbara
Santa Barbara, CA 93106

ABSTRACT

A survey is presented concerning fixed domain methods used to solve mathematical models of free and moving boundary flow problems in porous media. These include the following: variational inequality or quasi-variational inequality formulations; general inequality formulations which have been set and solved in fixed domains; and the residual flow procedure. Finally, some parallel computing methods and mesh adaptation methods are discussed to demonstrate how these fixed domain formulations can be solved with current technology.

The fixed domain methods that are referenced herein can be classified into two groups: the variational inequality method and the extended pressure head method. Baiocchi was the first to apply the variational inequality method to free boundary problems of flows through porous media. This method in general also uses an extension of the pressure head but adds an application of an integral transformation (a Baiocchi transformation) to the problem. The method possesses a beautiful mathematical structure for its theory and yields simple numerical solution algorithms. However, application of the method is difficult if not impossible in some cases depending upon the regularity of the seepage domain.

The extended pressure head method is based on the concept that the pressure is extended "smoothly" across the free or moving boundary into the unsaturated region from the flow domain. The extension of the pressure head to the entire porous medium yields an extended coefficient of permeability of the medium which is equal to the saturated coefficient in the seepage region and is equal to zero or some small value (for computational purposes) in the unsaturated region.

Key words. Fixed domain method, free boundary, moving boundary, Baiocchi transformation.

1 .INTRODUCTION

Flow through porous media has great importance in many facets of engineering practice. In particular such flows with free or moving boundaries make up a significant part of the seepage phenomena that occur in nature. Some examples of these are seepage through earth dams; seepage from open channels such as rivers, ponds, irrigation systems, and recharge basins; unconfined aquifer flow; and flow to wells.

In the analysis of seepage through porous media in the presence of a free or moving boundary, numerous solution approaches have been used (Bear, 1972). In addition, numerical schemes using the finite element or finite difference techniques have extended the solvable types of problems to a wider range of applications. For some examples see the references in Cryer (1976a), (1976b), and (1977).

Herein, a discussion is presented of several fixed domain formulations of these free and moving boundary seepage problems. They will be classified into the three categories: a) variational and quasi-variational inequality methods; b) general inequality methods; and c) residual flow procedure. Bruch (1989) presents a bibliography of research publications on these fixed domain methods. References to numerous applications are given in addition to relevant books, surveys and proceedings. Further, works by Westbrook (1985) and Desai and Baseghi (1989) present numerical comparisons of some of these methods. Westbrook (1985) shows that the general inequality method and the residual flow procedure have much in common when the numerical schemes resulting from these views are examined. He also makes some comparisons of numerical results obtained using the general inequality method and the variational inequality method applied to several seepage problems. Desai and Baseghi (1989) compare their numerical results using the residual flow procedure with the results of Caffrey and Bruch (1979) and Lin et al. (1984) who use the variational inequality method on a three-dimensional steady state seepage and a two-dimensional unsteady seepage through a dam, respectively.

In a paper in 1971, Baiocchi (1971) showed that the steady-state dam seepage problem for the homogeneous rectangular dam could be formulated as a variational inequality problem. In this approach introduced by Baiocchi (1971) and (1972) the dependent variable in the *a priori* unknown solution region was extended continuously across the free surface into a known region. A new dependent variable was then defined using Baiocchi's transformation within the new solution region which was the union of these two regions. This approach yielded a 'complementarity system' associated with a variational inequality which could be formulated. With this approach, theoretical results (uniqueness and existence), rigorous from a mathematical point of view, could be obtained. Also new numerical schemes on a fixed mesh were obtained. The resulting discrete problem was a quadratic programming problem. Comprehensive reviews of variational inequality methods for free surface seepage problems are also given by Oden and Kikuchi (1980) and Bruch (1980) and (1989).

In numerous subsequent papers, Baiocchi and his co-workers (Baiocchi et al. (1973a); Baiocchi et al. (1973b); see also Bruch (1989) for a listing of other work) gave variational inequality formulations for homogeneous dams with other geometries requiring more complicated formulations and numerical algorithms. They noted that for inhomogeneous dams a quasi-variational inequality formulation was required. Thus, by means of the Baiocchi transformation, free boundary seepage problems can be reduced to one variational inequality in the simplest cases; into a family of variational inequalities, depending upon one or more real parameters, in more complicated cases; and into a quasi-variational inequality in the general case. Extensions of the approach have been applied to steady-state three-dimensional problems and to unsteady seepage problems in two- and three-dimensions (Bruch, 1989).

For the more theoretical aspects of variational inequalities in general Mosco (1973) gives a good bibliography of work done on variational inequalities through 1973. One of the earliest and more significant works on theoretical questions concerning variational inequalities is the paper by Lions and Stampacchia (1967). In this paper they are concerned with questions of uniqueness, existence and continuous dependence of the solution on data. Further theoretical results concerning the continuity of the solution of variational inequalities and free boundary were obtained by Lewy and Stampacchia (1969) and (1971). Other results have been obtained by Kinderlehrer (1976), Friedman (1976a) and (1976b), Friedman and Jensen (1977), Caffarelli (1976), and Alt (1977a) and (1977b) to mention a few researchers. Applications of variational inequalities to many physical problems have been considered in the book by Duvaut and Lions (1976). In addition, the work by Glowinski et al. (1976) presents the optimization algorithms relevant to solving variational inequalities in addition to several applications including time dependent problems. Finally,

Oden and Kikuchi (1980) present a rather detailed survey of the theory of variational inequalities, their approximation and numerical analysis, and demonstrate applications of these theories to the analysis of difficult free boundary problems encountered in the study of flow through porous media.

In 1979 Alt presented a more general inequality formulation for the dam problem which could be applied to general geometry and to inhomogeneous problems. Using this formulation he obtained existence results and also showed the possibility of obtaining non-unique solutions. He went on to propose numerical schemes based on his formulation. He applied these to a variety of problems whose solutions may or may not be unique. Further, he includes proofs of the convergence of his numerical scheme to a solution, usually both a maximal and a minimal solution, of the problem.

Desai (1976) took a different view of these free and moving boundary seepage problems and introduced a residual flow procedure which again allows the use of a fixed mesh. A more solid formulation of the general ideas of this paper is given in Bathe and Khoshgoftaar (1979) and Desai and Li (1983), where numerical procedures and examples are also given. The formulation is heuristic for which convergence results are not available. However, from the numerical examples it is shown to be very effective.

The various fixed domain methods discussed will now be applied to the problem of steady-state seepage through a two-dimensional dam in order to demonstrate the approaches. The applicability of the approaches, however, extends well beyond this example (Bruch, 1989). For example, all the approaches can be applied to two-dimensional unsteady seepage problems or three-dimensional steady-state seepage problems.

2. VARIATIONAL AND QUASI-VARIATIONAL INEQUALITY METHODS

2.1 DIFFERENTIAL FORMULATION OF A MODEL PROBLEM -- SEEPAGE THROUGH A TWO-DIMENSIONAL DAM

The problem of steady flow of an incompressible fluid through a straight sided homogeneous dam on an impervious foundation is considered (Bruch, 1980). Figure 2.1 shows this dam which is separating two reservoirs of water heights y_1 and y_2, respectively. For this problem x_1 (the length of the dam), y_1 and y_2 are given numbers such that $x_1 > 0$, $y_1 > y_2 \geq 0$. In order to simplify the problem for discussion purposes several common assumptions are made. First, the soil in the flowfield is taken to be homogeneous and isotropic (k, the permeability of the soil, is constant). Secondly, capillary and evaporation effects are neglected. Thirdly, the flow is assumed to follow Darcy's law.

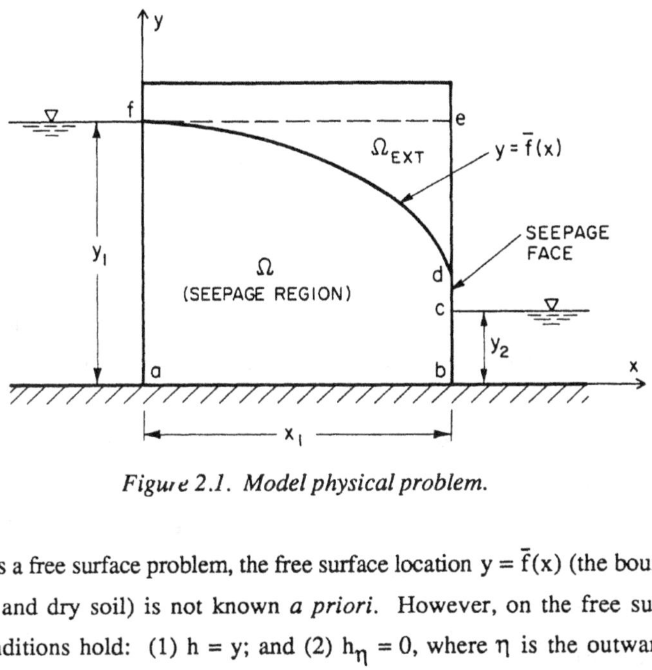

Figure 2.1. Model physical problem.

Since this is a free surface problem, the free surface location $y = \bar{f}(x)$ (the boundary line between wet and dry soil) is not known *a priori*. However, on the free surface two boundary conditions hold: (1) h = y; and (2) $h_\eta = 0$, where η is the outward normal direction and the piezometric head $h = (p/\rho g) + y$, in which p is the pressure, ρ is the density of the fluid, g is the gravitational acceleration, and y is the vertical coordinate (positive upward). On the remainder of the boundaries either Neumann or Dirichlet data exist. Also since some fluid emerges from the porous medium at the seepage face dc, the following condition $h_\eta \leq 0$ holds there.

The statement of the problem is to find $y = \bar{f}(x)$ such that: $\bar{f}(x)$ is defined and smooth in $[0, x_1]$; $\bar{f}(0) = y_1$, $\frac{d}{dx}\bar{f}(0) = 0$, $\bar{f}(x_1) > y_2$, $\frac{d}{dx}\bar{f}(x_1) = -\infty$. Setting $\Omega = \{(x,y) \mid 0 < x < x_1, 0 < y < \bar{f}(x)\}$, there exists a function h(x, y) defined and smooth in $\overline{\Omega}$ such that:

$$\nabla^2 h = 0\,, \qquad \text{in } \Omega\,, \tag{2.1}$$
$$h = y_1\,, \qquad \text{on } [af]\,, \tag{2.1a}$$
$$h = y_2\,, \qquad \text{on } [bc]\,, \tag{2.1b}$$
$$h = y\,, \qquad \text{on } [cd]\,, \tag{2.1c}$$
$$h = y\,, \qquad \text{on } \widehat{fd}\,, \tag{2.1d}$$
$$h_\eta = 0\,, \qquad \text{on } \widehat{fd}\,, \tag{2.1e}$$
$$h_y = 0\,, \qquad \text{on } [ab]\,, \tag{2.1f}$$
$$h_\eta = \leq 0\,, \qquad \text{on }]cd[\,. \tag{2.1g}$$

Jensen (1977) and Alt (1977b) have shown the analyticity of the free boundary, $\bar{f}(x)$.

A standard numerical method (Cryer, 1976b) to solve this type of problem has been to assume a position for the free surface which means that the unknown flow domain is known *a priori*. Then this trial domain is corrected through an iterative technique such as checking to see if the normal velocity of the flow along the free surface is equal to zero. If it is not, the free surface is moved to a new position and the problem is resolved. This is done until the normal velocity is within some error criterion.

A weak solution to the problem given by Equations (2.1) is a triplet $\{\bar{f}, \Omega, h\}$ such that: $\bar{f}(x)$ is a continuous strictly decreasing function on $[0, x_1]$ with the previously given properties; Ω is the open set defined previously; $h(x, y) \in H^1(\Omega) \cap C^0(\Omega)$ and satisfies Equations (2.1a) to (2.1d) in the sense of $C^0(\overline{\Omega})$; and if $\bar{\varphi}(x,y) \in C^1([0,x_1]\times[0,y_1])$ with $\bar{\varphi} = 0$ on a neighborhood of $[af] \cup [bd]$:

$$\iint_\Omega (h_x \bar{\varphi}_x + h_y \bar{\varphi}_y)\, dxdy = 0\,. \tag{2.2}$$

2.2 FIXED DOMAIN FORMULATION

The method discussed now to solve this free surface problem is the Baiocchi method and transformation (Baiocchi, 1971 and 1972). Let the solution domain Ω be extended to the known region: $D = \{(x, y) \mid 0 < x < x_1, 0 < y < y_1\}$ in Figure 2.1 where $D = \Omega + \Omega_{ext} + \Gamma_{\widehat{fd}}$, $\Omega_{ext} = \{(x, y) \mid 0 < x < x_1,\ \bar{f}(x) < y < y_1\}$, and $\Gamma_{\widehat{fd}} = \{(x, y) \mid 0 < x < x_1,\ y = \bar{f}(x)\}$. Then extend h continuously to be defined on $\overline{D}$ by setting:

$$\tilde{h}(x\,y) = h(x,y)\,, \qquad \text{in } \overline{\Omega} \tag{2.3a}$$

and

$$\tilde{h}(x,y) = y\,, \quad \text{in } \overline{D} - \overline{\Omega}\,. \tag{2.3b}$$

Using Equations (2.3) it can be shown that $\tilde{h}$ satisfies the equation $-\nabla^2\tilde{h} = (\chi_\Omega)_y$ in the sense of distributions on D, where χ_Ω is the characteristic function ($\chi_\Omega = 1$ in Ω, $\chi_\Omega = 0$ in $D - \overline{\Omega}$).

Next a new dependent variable w is defined using the Baiocchi transformation:

$$w(x,y) = \int_y^{y_1} [\tilde{h}(x,\overline{\eta}) - \overline{\eta}]\, d\overline{\eta}\,. \tag{2.4}$$

The general idea of the Baiocchi transformation is to obtain a formulation for the extended problem in which the derivative of the characteristic function term does not appear.

In the sense of distributions the governing differential equation in D can be obtained using Equation (2.4). Thus,

$$\nabla^2 w(x,y) = \chi_\Omega\,, \quad \text{in } D \tag{2.5}$$

for all $w(x,y) \in H^2(D) \cap C^1(\overline{D})$, i.e., $\nabla^2 w(x,y) = 1$ when $w > 0$ and $\nabla^2 w(x,y) = 0$ when $w = 0$. Hence, in D, $w(x,y) \geq 0$; $1 - \nabla^2 w(x,y) \geq 0$; and $w(1 - \nabla^2 w) = 0$. The latter equations are the variational inequality (which is given in Equation (2.7)) written in 'complementarity' form. From Equation (2.5) it follows that $0 \leq \nabla^2 w \leq 1$.

To complete the problem boundary conditions are needed. These are:

$$w(0,y) = \frac{1}{2}(y_1 - y)^2\,, \quad \text{on [af]}\,, \tag{2.5a}$$

$$w(x_1,y) = \frac{1}{2}(y_2 - y)^2\,, \quad \text{on [bc]}\,, \tag{2.5b}$$

$$w(x_1,y) = 0\,, \quad \text{on [cd]}\,, \tag{2.5c}$$

$$w(x_1,y) = 0\,, \quad \text{on [de]}\,, \tag{2.5d}$$

$$w(x,y_1) = 0\,, \quad \text{on [fe]}\,, \tag{2.5e}$$

$$w(x,0) = \frac{y_1^2}{2} - \frac{y_1^2 - y_2^2}{2x_1}\,x\,, \quad \text{on [ab]}\,. \tag{2.5f}$$

If a w satisfying Equations (2.5) is found, then the following quantities can be obtained:

$$\Omega = \{(x,y) \mid (x,y) \in D,\ w(x,y) > 0\}\,, \tag{2.6a}$$

$$\text{graph } \bar{f} = \partial\Omega - \partial D \equiv \text{points of } \partial\Omega \text{ not on } \partial D \tag{2.6b}$$

or

$$\bar{f}(x) = \sup\{y \mid (x,y) \in \Omega\}, \quad 0 < x < x_1\,,$$

with

$$\bar{f}(0) = \lim_{x\to 0^+} \bar{f}(x) \quad \text{and} \quad \bar{f}(x_1) = \lim_{x\to x_1^-} \bar{f}(x)\,, \tag{2.6c}$$

$$h(x,y) = y - w_y(x,y) \text{ restricted to } \overline{\Omega}\ . \tag{2.6d}$$

2.3 VARIATIONAL INEQUALITY FORMULATION

Let $\overline{w}(x,y)$ be the Dirichlet data, Equations (2.5a)-(2.5f), on ∂D. Define $K = \{v | v \in H^1(D),\ v|_{\partial D} = \overline{w},\ v \geq 0$ a.e. on $D\}$, a closed convex set, $K \subset H^1(D)$.

Theorem 1: If $w \in K$ satisfies Equations (2.5), then it also satisfies the variational inequality:

$$a(w,\ v - w) \geq L(v - w), \quad \forall v \in K, \tag{2.7}$$

where

$$a(w,\ v - w) = \iint_D \nabla w \cdot \nabla(v - w)\, dxdy$$
$$= \iint_D \{w_x(v_x - w_x) + w_y(v_y - w_y)\}\, dxdy\,, \tag{2.7a}$$

$$L(v - w) = -\iint_D (v - w)\, dxdy\ . \tag{2.7b}$$

The converse of Theorem 1 also holds. If $w \in K$ satisfies Equation (2.7) for all $v \in K$, then it also satisfies Equations (2.5).

It should further be noted that because of the symmetry in Equation (2.7), the finding of $w \in K$ is equivalent to solving the minimization problem:

$$J(w) \leq J(v), \quad \forall v \in K\,, \tag{2.8}$$

where

$$J(v) = (1/2) \iint_D [(v_x)^2 + (v_y)^2]\, dxdy + \iint_D v dxdy\ . \tag{2.8a}$$

The uniqueness of the solution of the problem, Equations (2.5), follows from the uniqueness of the solution of Equation (2.7).

Theorem 2: The variational inequality, Equation (2.7), has at most one solution $w \in K$.

In regards to the existence of such solutions the following two theorems will be stated. See Baiocchi et al. (1973b) and (1976b), Lions and Stampacchia (1967) and Glowinski (1976) for proofs of these theorems.

Theorem 3: There exists one $w \in K$ which is a solution of Equation (2.7).

Theorem 4: Let w be a solution of Equation (2.7) and let Ω, $\bar{f}$, and h be defined by

Equations (2.6a), (2.6b) and (2.6d); then $\{\bar{f}, \Omega, h\}$ is a solution of Equations (2.5).

Another type of Baiocchi transformation that could have been used here (instead of Equation (2.4)) is one which is a line integral and includes both the velocity potential and stream function (Bruch, 1980). This integral transformation extends the applicability of the approach to more general geometries which can be solved numerically very simply using the 'complementarity system' but presents theoretical complications when trying to obtain the variational inequality formulation.

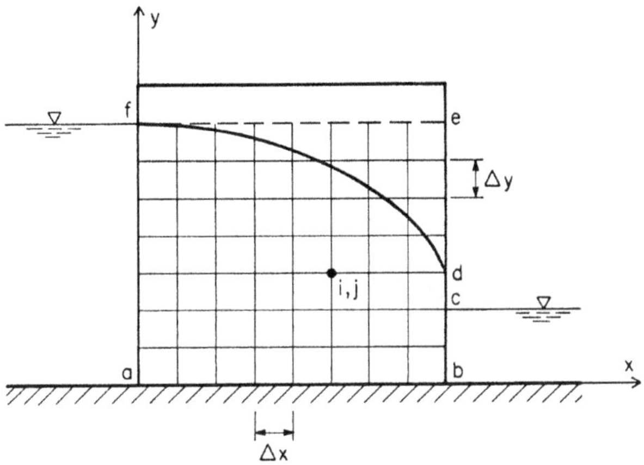

Figure 2.2. Mesh on solution region for numerical algorithms.

For the model problem (Figure 2.1) being discussed the solution domain is the rectangle abef and simple numerical algorithms for the problem are given in Baiocchi et al. (1973a) and Bruch (1980). Thus, the computer programming effort involved in applying algorithms is very minimal. Furthermore, the approximate location of the free surface can be picked off very easily from the computer printout by noting the first zero that occurs in moving up the mesh in the y-direction for a fixed x-value on the mesh (see Figure 2.2). This would give a discontinuous approximate surface through which a smooth curve could be placed. By interpolating appropriately along the free surface a more accurate shape would be obtained. Finally, decreasing the mesh size smooths out the free surface and there would be convergence to the actual free surface in the limit as the mesh size goes to zero. Baiocchi et al. (1973a) give numerical results for the model problem for a set of geometrical and discretization parameters. See Bruch (1989) for references to papers having numerical results to numerous other examples.

2.4 QUASI-VARIATIONAL INEQUALITY FORMULATION

The physical problem to be solved is the steady-state flow through the arbitrary shaped earth dam shown in Figure 2.3 separating two water reservoirs of different heights (Baiocchi, 1975a). The medium is assumed to be homogeneous and isotropic for this formulation. The cross-section of the dam has the form: $D = \{(x,y) \mid 0 < x < x_1;\ 0 < y < Y(x)\}$, where Y is a smooth function on $0 < x < x_1$ with $Y_{xx}(x) \leq 0$, i.e. a concave function; $Y(x_f) = y_1$ (the height of the reservoir on the left) and $Y(x_c) = y_2$ (the height of the reservoir on the right); and x_f and x_c are the horizontal distances along the x-axis to points f and c, respectively.

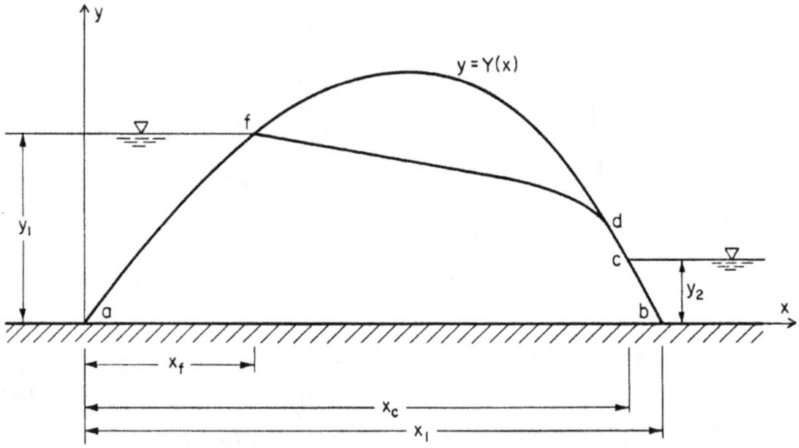

Figure 2.3. Seepage through an arbitrary shaped dam.

Let Ω be the flow region defined as: $\Omega = D_1 \cup D_2 \cup \{(x,y) \mid (x,y) \in D_3\ ,\ y < \tilde{f}(x)\}$, where $D_1 = \{(x,y) \mid (x,y) \in D,\ 0 < x \leq x_f\}$; $D_2 = \{(x,y) \mid (x,y) \in D,\ x_c \leq x \leq x_1\}$; $D_3 = \{(x,y) \mid (x,y) \in D,\ x_f < x \leq x_c\}$; and $\tilde{f}(x) = \sup\{y \mid (x,y) \in \Omega\}$ for $x_f \leq x \leq x_c$. The mathematical model for this problem is given by Equations (2.1) where now $\widehat{af}$, $\widehat{bc}$, and $\widehat{cd}$ are arbitrary curves.

Extending h as before into $\overline{D} - \overline{\Omega}$ using Equations (2.3) and using the change of variable

$$z(x,y) = \int_0^y [\bar{\eta} - \tilde{h}(x,\bar{\eta})]\, d\bar{\eta} , \qquad \text{in } \bar{D} \tag{2.10}$$

yields the following problem for $z(x,y) \in H^2(D) \cap C^1(\bar{D})$:

$$\nabla^2 z = \chi_\Omega , \text{ in } D\ (0 \le \nabla^2 z \le 1 \text{ in } D), \tag{2.11}$$

$$z(x,y) \ge z(x, Y(x)), \quad 0 \le x \le x_1, \quad 0 \le y \le Y(x), \tag{2.11a}$$

$$z = 0 , \qquad \text{on } [ab] , \tag{2.11b}$$

$$z_y = y - y_1 , \qquad \text{on } \widehat{af} , \tag{2.11c}$$

$$z_y = y - y_2 , \qquad \text{on } \widehat{bc} , \tag{2.11d}$$

$$z_y = 0, \qquad \text{on } y = Y(x),\ x_f < x < x_c , \tag{2.11e}$$

and $z(x, Y(x))$ is concave as a function of x in $[x_f, x_c]$. Equation (2.11) could also be written:

$$\nabla^2 z \in \chi_{D_1} + \chi_{D_2} + \chi_{D_3} H[z - z(x, Y(x))] , \tag{2.12}$$

where χ_{D_i} is the characteristic function of D_i and H is the maximal monotonic graph associated with the Heaviside function.

From the solution of Equations (2.11) the parameters $\{\Omega, h, \tilde{f}(x), q\}$ can be obtained as follows:

$$\Omega = D_1 \cup D_2 \cup \{(x,y) | (x,y) \in D_3,\ z(x,y) > z(x, Y(x))\} , \tag{2.13a}$$

$$h = y - z_y(x,y) , \quad \text{in } \bar{\Omega} , \tag{2.13b}$$

$$\tilde{f}(x) = \sup\{y \mid (x,y) \in \Omega\} , \quad \text{for } x_f \le x \le x_c , \tag{2.13c}$$

$$q = z_x(x, Y(x)) , \quad \text{for } x_f \le x \le x_c , \tag{2.13d}$$

where q is the flowrate through the flowfield.

The problem stated in Equations (2.11) can be reformulated into the following problem: $z \in H^1(D)$; $z(x,y) = 0$ on [ab]; for any $v \in H^1(D)$ with $v(x,y) = 0$ on [ab]

$$a(z, v - z) + j(z, v) \ge j(z, z) + L(v - z) , \tag{2.14}$$

where

$$a(z, v - z) = \iint_D [(z_x - Y_x z_y)(v - z)_x + (z_y + Y_x z_x)(v - z)_y - Y_{xx}\, z_y (v - z)]\, dxdy , \tag{2.14a}$$

$$L(v - z) = -\iint_{D_1 \cup D_2} (v - z)\, dxdy + \int_0^{x_f} (1+Y_x^2)\,[Y(x) - y_1]\,[v(x, Y(x)) - z(x, Y(x))]\, dx + \int_{x_c}^{x_1} (1+Y_x^2)\,[Y(x) - y_2]\,[v(x, Y(x)) - z(x, Y(x))]\, dx\,, \tag{2.14b}$$

$$j(z, v) = \iint_{D_3} [v(x,y) - z(x, Y(x))]^+\, dxdy\,, \tag{2.14c}$$

$$j(z, z) = \iint_{D_3} [z(x,y) - z(x, Y(x))]^+\, dxdy\,, \tag{2.14d}$$

in which $t^+ = t$ for $t \geq 0$ and $t^+ = 0$ for $t \leq 0$.

The problem shown in Equations (2.14) to solve for z is of the type called 'quasi-variational inequality', i.e. the function z(x, Y(x)) (see Equation (2.12)) depends itself on the unknown function z. Quasi-variational inequalities were introduced by Bensoussan and Lions (1973) in order to study some impulse control problems. The quasi-variational approach of Baiocchi given above runs into difficulties in uniqueness and regularity.

For problems similar to that given by the previous theory Baiocchi (1975a), Baiocchi et al. (1975) and Comincioli (1974) demonstrate the existence of a maximal solution, z_{max}, and a minimal solution, z_{min}. Furthermore, Comincioli (1974) and Baiocchi (1975a) present efficient and rigorous algorithms to obtain approximations of z_{max} and z_{min}. Numerical experiments showed that $z_{min} = z_{max}$, so they have conjectured uniqueness for their quasi-variational inequalities (Baiocchi, 1975b; Baiocchi et al., 1976a; and Gilardi, 1976).

Some of the attributes and advantages of the variational and quasivariational methods are: (1) they involve invariant meshes which avoids the need for mesh modification; (2) they have rigorous mathematical bases; (3) they yield efficient numerical algorithms; and (4) they can take into account a capillary fringe (Comincioli and Guerri, 1976). Some of the disadvantages of these methods are: (1) an integral transformation is needed; (2) there is no general formulation for problems with arbitrary geometries and permeabilities (different geometry and boundary conditions may lead to different formulations); and (3) the necessity to transform the pressure variable makes the procedure awkward to couple with the required solution procedure for stress analysis in the soil.

3. GENERAL INEQUALITY METHOD

In the late 1970's Brezis et al. (1978) and Alt (1979), (1980a) and (1980b) introduced a new approach and formulation which required no integral transformation. It was a direct method for resolving the seepage problem which numerically competes well with the method of variational inequalities. This formulation was more complicated, the solution less regular, but it allowed more general cases to be handled especially in regard to the geometry of the domain. In addition Alt's formulation allowed the possibility of partially saturated flowfields. Also this general formulation was treated from a numerical point of view by Alt (1980a) and (1980b). However, in the framework of his assumptions he was restricted in the type of finite elements he could use in his discretization. Thus fixed boundaries could not be modeled precisely. Nevertheless, he was able for steady state cases to predict zones of saturation and partial saturation in seepage flowfields. This is a drawback of the Baiocchi method, as seen in Baiocchi and Friedman (1977) and Remar et al. (1984). In the following discussion of the general inequality method Alt's notation will be used in general and Figure 2.3 again depicts the flow problem.

The porous medium is again denoted by D, where D is a bounded, connected open set of R^n with boundary ∂D. The boundary ∂D has three parts: $S^0 = \{(x,y) \mid x_f < x < x_c,\ y = Y(x)\}$ is the boundary with the atmosphere, $S^+ = \widehat{af} \cup \widehat{cb}$ is in contact with water, and $\partial D - S^0 - S^+ = [ab]$ is pervious where the flux is prescribed by a function f^0 to allow for a leaky region in the base of the dam. If this latter part of the boundary is impermeable, $f^0 = 0$ there. It is assumed that S^0 and S^+ are measurable and disjoint sets and that the measure of S^+ is positive. The medium is assumed to be non-homogeneous and anisotropic. Therefore, the permeability is given by a matrix function K, where K is strictly positive definite. The boundary data are given by a function p^0 which is non-negative and together with its first and second derivatives square summable on D. The following problem is now considered:

Find a pair $(p, \beta) \in H^1(D) \times L^\infty(D)$ such that $p \geq 0$ a.e. in D, $p = p^0$ (hydrostatic pressure) on S^+, $p \leq p^0$ on S^0; $0 \leq \beta \leq 1$ a.e. in D, $\beta = 1$ a.e. on $\{p > 0\}$;

$$\int_D \nabla(v-p)\cdot K(\nabla p + \beta\vec{e})\, dD - \int_{\partial D}(v-p)f^0 dx \geq 0\ , \qquad \forall v \in W\ , \tag{3.1}$$

where $W = \{\overline{w} \in H^1(D) \mid \overline{w} = p^0 \text{ on } S^+,\ \overline{w} \leq p^0 \text{ on } S^0\}$ and $\vec{e}$ is the vertical unit vector in the direction of gravity.

The assumption is that unsaturated flow is controlled by an extension of Darcy's Law, i.e.

$$\vec{q} = -K(\nabla p + \beta \vec{e}), \tag{3.2}$$

so that the inequality is formally equivalent to the differential equation

$$\nabla \cdot \vec{q} = 0, \quad \text{in } D \tag{3.3}$$

and the boundary conditions

$$p = p^0, \quad \text{on } S^+, \tag{3.4a}$$

$$\vec{q} \cdot \vec{\eta} + f^0 = 0, \quad \text{on} \quad \partial D - S^0 - S^+, \tag{3.4b}$$

$$p \le p^0 \text{ and } \vec{q} \cdot \vec{\eta} \ge 0 \text{ and } (p - p^0)\, \vec{q} \cdot \vec{\eta} = 0, \quad \text{on } S^0, \tag{3.4c}$$

where $\vec{q}$ is the velocity of the porous flow and $\vec{\eta}$ is the unit outward normal vector to the relevant surface. Furthermore, the pressure is normalized, such that the capillary pressure of the porous medium is zero, that is, the atmospheric pressure may be strictly positive.

The function β is here a nonlinear function of the degree of saturation, which is characteristic of the porous medium (Bear 1972, p. 496). The set $\{\beta = 0\}$ is the dry region while the set $\{\beta = 1\}$ is the saturated region. If the set $\{0 < \beta < 1\}$ has a positive Lebesgue measure, the unique solution has an unsaturated region (such as a capillary fringe, see Alt (1979), (1980a) and (1980b) for an example of this effect). Alt (1979) proved an existence result for p and β in Equation (3.1).

In a homogeneous medium the solution consists only of saturated regions where $p > 0$ and $\beta = 1$ and dry regions where $\beta = 0$. Alt (1979) proved that the inequality $\nabla \cdot (K\vec{e}) \ge 0$ is always satisfied in these cases. The function p is the solution of an elliptic partial differential equation in the saturated regions. However, for general inhomogeneous media the condition $\nabla \cdot (K\vec{e}) < 0$ may hold somewhere and then an unsaturated region may exist in which $0 < \beta < 1$. In this region $p = 0$ and the flow is determined by the gravitational force and not by the pressure gradient. The function β is a solution of a first-order differential equation of flow. Alt (1979) analyzes a simple example of an unsaturated region in a horizontally stratified inhomogeneous medium. The physical significance of the condition $\nabla \cdot (K\vec{e}) < 0$ is that it permits the continuity of flow condition to be satisfied between a saturated and an unsaturated region.

Some of the attributes and advantages of the general inequality method are: (1) it involves invariant meshes which avoids the need of mesh modification; (2) it is able to handle general geometries; (3) it can accommodate inhomogeneous and anisotropic media; and (4) it allows the possibility of partially saturated regions. Some of the disadvantages are: (1) it is a complicated formulation; (2) its solution is less regular; and (3) the solution domain boundaries have to be approximated (a restriction on types of elements used). See Pietra (1982) and Marini and Pietra (1986) for ways of handling this latter difficulty.

4 .RESIDUAL FLOW PROCEDURE

A different view of these free and moving boundary seepage problems was taken by Desai (1976) when he introduced the method of residual flow or Residual Flow Procedure (RFP). The method uses a fixed domain. The procedure will be presented in conjunction with the finite element method in order to see the essence of the approach. Again the notation used by Desai and his co-workers will in general be used as will the flow problem shown in Figure 2.3.

For incompressible fluid flow through a porous medium such as the flow domain (dam) shown in Figure 2.3, the governing equation of flow can be written as

$$\operatorname{div}(k \operatorname{grad} h) = 0 \ . \tag{4.1}$$

Equation (4.1) is assumed to hold in both Ω and D-Ω by introducing a definition of k (the coefficient of permeability) as follows:

$$k(p) = \begin{cases} k_s, & \text{in } \Omega \\ k_{us} = k_s - f(p), & \text{in } D - \Omega, \end{cases} \tag{4.2}$$

where subscripts 's' and 'us' denote saturated and unsaturated zones, respectively, and f(p) is a function of pressure p. Typical pressure-permeability relations are depicted in Figure 4.1. If such relations are not available from laboratory and/or field tests (results shown in the solid line representation), for the implementation of the RFP it can be appropriate to assume simplified relations such as the dotted line example shown.

Equation (4.1) with k defined as in Equation (4.2) may be regarded as similar to that used for saturated-unsaturated flow problems (Bouwer, 1964). The value of k in the extended domain D - Ω can often be defined as a very small constant value (Bathe and Khoshgoftaar, 1979):

$$k(p) = \begin{cases} k_s, & \text{in } \Omega(p \geq 0) \\ k_s/1000, & \text{in } D - \Omega\,(p < 0). \end{cases} \tag{4.3}$$

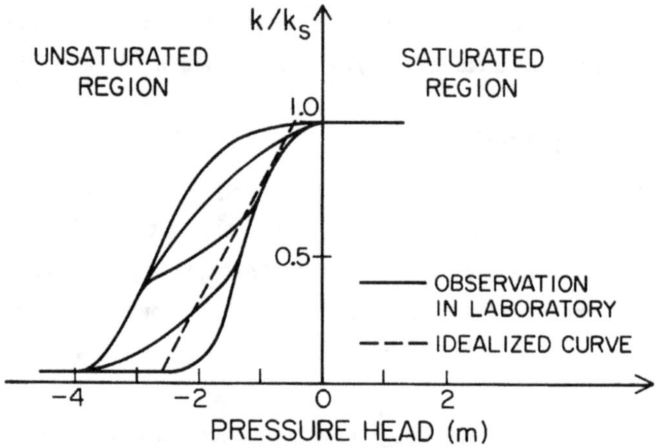

Figure 4.1. Pressure-Permeability Curves.

Desai and Li (1983) incorporate the linear permeability-pressure curve into their proposed finite element model. Such an assumption is considered appropriate in the case of most earth dams. See also Rank and Werner (1986).

The associated boundary conditions are expressed as

$$h = y_1 , \qquad \text{on } \widehat{af} , \tag{4.3a}$$

$$h = y_2 , \qquad \text{on } \widehat{bc} , \tag{4.3b}$$

$$k_x \frac{\partial h}{\partial x} l_x + k_y \frac{\partial h}{\partial y} l_y = \overline{q}_n , \qquad \text{on } [ab] = \Gamma_2 , \tag{4.3c}$$

$$h = y , \qquad \text{on } \widehat{cd} , \tag{4.3d}$$

$$k_x \frac{\partial h}{\partial x} l_x + k_y \frac{\partial h}{\partial y} l_y < 0 , \qquad \text{on } \widehat{cd} , \tag{4.3e}$$

$$h = y , \qquad \text{on } \widehat{fd} , \tag{4.3f}$$

$$k_x \frac{\partial h}{\partial x} l_x + k_y \frac{\partial h}{\partial y} l_y = 0 , \qquad \text{on } \widehat{fd} , \tag{4.3g}$$

where $\overline{q}_n$ is the specified intensity of flow. If $\overline{q}_n = 0$, the base is impermeable. Here, k_x and k_y are the coefficients of permeabilities in x and y directions, respectively, and l_x, l_y

are the direction cosines of outward normal to the boundaries.

The functional, U, for this seepage case can be expressed as

$$U(h) = \frac{1}{2}\iint_D \left\{(k_s - f)\left[\left(\frac{\partial h}{\partial x}\right)^2 + \left(\frac{\partial h}{\partial y}\right)^2\right]\right\} dD - \int_{\Gamma_2} \bar{q}_n h \, dx . \tag{4.4}$$

By taking the variation of the function in Equation (4.4), finding the stationary value and using the finite element method, the following element equations can be obtained:

$$[k_s]^e\{H\}^e - [k_{us}]^e\{H\}^e = \{Q\}^e \tag{4.5a}$$

or

$$[k_s]^e\{H\}^e = \{Q\}^e + [k_{us}]^e\{H\}^e = \{Q\}^e + \{Q_r\}^e, \tag{4.5b}$$

where

$$[k_s]^e = \iint_{D^e} [B]^T[R]\{B\}\, dD^e ,$$

$$[k_{us}]^e = \iint_{D^e} [B]^T[f][B]\, dD^e ,$$

$$\{Q\}^e = \int_{\Gamma_2^e} \{N\}^T\{\bar{q}_n\}\, dx$$

in which $\{H\}^e$ = vector of element nodal heads; $\{N\}$ = vector of interpolation functions (shape functions); $\{Q\}^e$ = vector of boundary element flux contributions; $\{Q_r\}^e$ = element residual flow vector; $[B]$ = gradient-head transformation matrix; $[R] = \begin{bmatrix} k_x & 0 \\ 0 & k_y \end{bmatrix}$ is the matrix of (principal) permeabilities; $[f] = \begin{bmatrix} f_x & 0 \\ 0 & f_y \end{bmatrix}$ is the matrix of (principal) values of f in Equation (4.2), D^e is the element area and Γ_2^e is the part of the element boundary that intersects Γ_2.

Since the free surface interface boundary is unknown *a priori* and the function f varies with the unknown h, solution of Equation (4.5b) can be difficult. Therefore, an iterative scheme is incorporated into the finite element procedure by assuming a fully saturated domain D in the first iteration. Then and after every iteration, based on the computed nodal heads, the separation of fully saturated and unsaturated zones, the intersection of free surface with the mesh, is found by linearly interpolating the pressure head between nodal points to locate points of zero pressure. The elements intersected by the free surface would usually contain both saturated and partially saturated zones. This influence on the element equations is introduced by considering contributions of only the unsaturated part in the computation of $[k_{us}]^e$. It is done by calculating the pressures at the integration points and by including contributions of points with $p > 0$ in $[k_s]^e$, and those of points with $p < 0$ in $[k_{us}]^e$. The negative values of pressure head ($p/\gamma = h - y$) in the unsaturated zone allows evaluation of $[k_{us}]$ in Equation (4.5b) from Equation (4.2), Figure 4.1 in order to form a residual flow vector where γ is the fluid density. The foregoing approach is used after each iteration in order to identify the elements in the unsaturated zone. It is also used to locate the free surface at the end of the iterative procedure. Thus the procedure based on an invariant mesh involves progressive correction of free surface solutions by using a residual or correction vector until convergence of the computed heads is achieved.

Some of the attributes and advantages of the RFP are: (1) it involves invariant meshes which avoids the need of mesh modifications; (2) it permits use of arbitrary shapes of (isoparametric) elements, whereas the general inequality methods may often be found difficult to implement for such shapes; (3) it can accommodate variable material properties whereas in the variable mesh schemes they can involve considerable difficulties; (4) it can incorporate partially saturated zones; (5) it is found to be relatively economical; e.g., in comparison with the inequality methods (see Desai and Baseghi, 1989, for these comparisons); and (6) it involves equations which are consistent with the general formulation for geotechnical problems. Some of the disadvantages of the RFP are: (1) a pressure-permeability relationship or data needed (idealized curves used) and (2) it is a heuristic formulation for which convergence results are not available.

5. PARALLEL COMPUTATION AND FINITE ELEMENT MESH ADAPTATION METHODS

Current computational technology that has recently been applied to the problem of seepage through two-dimensional dams includes the use of parallel algorithms in concurrent processor systems and the use of adaptive meshes in conjunction with the finite element method. White (1987) discusses parallel versions of the Gauss-Seidel SOR algorithm. The three basic parallel methods he describes are coloring, domain decomposition (or substructuring), and overlapping blocks (or multisplitting). He obtains numerical results for the problem of steady-state fluid flow in a porous medium (seepage through a two-dimensional rectangular homogeneous dam with reservoirs on both sides) using these various parallel methods on an Alliant FX/8.

Wang and Bruch (1989) present results of using parallel algorithms for solving a steady, two-dimensional free surface seepage through a slanted faced homogeneous dam with a toe drain on an Intel iPSC 32 node Hypercube Concurrent Computer. Projected successive over-relaxation (SOR) and Jacobi iteration schemes were used to solve the finite difference equations. In addition, the solution domain was split into two overlapping subdomains whose governing equations are in terms of different but related dependent variables. Therefore, data can be interchanged in the overlapping area. Also, during the computation, these two overlapping subdomains were divided again into a certain number of overlapping subdomains in order to distribute the computation tasks into processors of the parallel computer. A modified alternating iteration scheme was then used. The performance and the feasibility of the Hypercube Concurrent Computer are discussed by comparison with the number of nodes used and with a VAX 11/780.

In regards to using mesh adaptation approaches, Rank and Werner (1986) consider a non-linear model problem, the two-dimensional fluid flow through a rectangular porous homogeneous dam with a toe drain, and show how results from linear *a posteriori* theory can be used to control the non-linear iteration and to refine the mesh adaptively. A numerical example shows that, similar to linear problems, considerable improvement of the accuracy is obtained by an adaptive mesh refinement and that the influence of singularities on the order of convergence disappears.

Chung and Kikuchi (1987) discuss grid adaptive methods combined with domain adaptation for two-dimensional seepage flow problems with free boundaries through porous media. They present examples of grid and domain adaptive methods to demonstrate several ways to predict grids and shapes of free boundaries using an iterative scheme. Finally, they apply the combined adaptive methods to obtain smooth non-oscillatory shape of a free boundary of seepage flow through non-homogeneous porous media.

Burkley (1989) in the second half of his M.S. Thesis presents numerical experiments in adapting variations of a computationally simple error estimator to an existing finite element code using free surface seepage problems as examples. The error estimator used allows both overall and local errors to be estimated. From the local estimates of error, refinements of the mesh are calculated to reach a prescribed error tolerance. These calculated refinements are used by a mesh refiner to produce a modified mesh which lowers the overall error to the prescribed value while keeping the mesh as crude as possible. Further, the numerical examples demonstrate how the problem formulation affects the error analysis and how the choice of error affects the mesh adaptation.

6. ACKNOWLEDGEMENT

This work was partially supported by the National Science Foundation under Grant No. INT 86-13994.

7. REFERENCES

Alt, H.W., 1977a, The fluid flow through porous media. Regularity of the free surface, Manuscripta Math., 21, 255-272.

Alt, H.W., 1977b, A free boundary problem associated with the flow of ground water, Arch. Rat. Mech. Anal., 64, 111-126.

Alt, H.W., 1979, Stromungen durch inhomogene porose Medien mit freiem Rand, J. Reine Angew. Math, 305, 89-115.

Alt, H.W., 1980a, Numerical solution of steady-state porous flow free boundary problems, Numer. Math., 36, 73-98.

Alt, H.W., 1980b, A new numerical method for solving the dam problem, Proc. Seminar on Free Boundary Problems, Pavia 1979, Ist. Naz. Alta Mat., Roma, 1, 89-108.

Baiocchi, C., 1971, Sur un probleme a frontiere libre traduisant le filtrage de liquides a travers des milieux poreux, C.R. Acad. Sci. Paris, 273, 1215-1217.

Baiocchi, C., 1972, Su un problema di frontiera libera connesso a questioni di idraulica, Ann. Math. Pura Appl, 92(4), 107-127.

Baiocchi, C., 1975a, Free boundary problems in the theory of fluid flow through porous media, Proc. Int. Congr. Math. Vancouver, Vancouver, 1974, II, 237-243.

Baiocchi, C., 1975b, Studio di un problema quasi-variazionale connesso a problemi di frontiera libera, Boll. UMI., 11, 589-613.

Baiocchi, C., Brezzi, F. and Comincioli, V., 1976a, Free boundary problems in fluid flow through porous media, Proc. 2nd Int. Symp. Finite Element Methods in Flow Problems, ICCAD, Santa Margherita Ligure, 409-420.

Baiocchi, C., Comincioli, V. and Maione, U., 1975, Unconfined flow through porous media, Meccanica-J., Ital. Assoc. Theoret. Appl. Mech., 10, 151-155.

Baiocchi, C., Comincioli, V., Guerri, L. and Volpi, G., 1973a, Free boundary problems in the theory of fluid flow through porous media: a numerical approach, Calcolo, 10, 1-86.

Baiocchi, C., Comincioli, V., Magenes, E. and Pozzi, G.A., 1973b, Free boundary problems in the theory of fluid flow through porous media: existence and uniqueness theorems, Ann. Mat. Pura Appl., 93(4), 1-82.

Baiocchi, C., Comincioli, V., Magenes, E. and Pozzi, G.A., 1976b, Fluid flow through porous media: a new theoretical and numerical approach, Proc. Int. Autumn Course, Math. Numerical Methods in Fluid Dynamics, Int. Centre for Theoretical Physics, Trieste, 395-447.

Baiocchi, C. and Friedman, A., 1977, A filtration problem in a porous medium with variable permeability, Ann. Mat. Pura Appl., 114(4), 377-393.

Bathe, K.J. and Khoshgoftaar, M.R., 1979, Finite element free surface seepage analysis without mesh iteration, Int. J. for Num. and Anal. Meth. in Geomech., 3, 13-22.

Bear, J., 1972, Dynamics of Fluids in Porous Media, American Elsevier, New York.

Bensoussan, A. and Lions, J.L., 1973, Nouvelle formulation de problemes de controle impulsionnel et applications, C.R. Acad. Sci. Paris, (A-B), 276, 1189-1192.

Bouwer, H., 1964, Unsaturated flow in ground-water hydraulics, J. of Hydraulics Div., ASCE, 90, 121-144.

Brezis, H., Kinderlehrer, D., and Stampacchia, G., 1978, Sur une nouvelle formulation du probleme de l'ecoulement a travers une digue, C.R. Acad. Sci. Paris, 287, 711-714.

Bruch, J.C., Jr., 1980, A survey of free boundary value problems in the theory of fluid flow through porous media: variational inequality approach, Adv. in Water Resources Part I, 3, 65-80; Part II, 3, 116-124.

Bruch, Jr., J.C., 1989, Bibliography of fixed domain methods in flow through porous media, Tech. Rept. UCSB-ME-89-5, Dept. of Mech. and Env. Engr., Univ. of Calif., Santa Barbara, Calif., 21 pp.

Burkley, V.J., 1989, Engine optimization and adaptive error analysis, M.S. Thesis, Dept. of Mech. and Envr. Engr., Univ. of Calif., Santa Barbara, Calif., 139 pp..

Caffarelli, L.A., 1976, The smoothness of the free surface in a filtration problem, Arch. Rat. Mech. Anal., 63, 77-86.

Caffrey, J. and Bruch, J.C., Jr., 1979, Three-dimensional seepage through a homogeneous dam, Advances in Water Resources, 2(4), 167-176.

Chung, K.Y. and Kikuchi, N., 1987, Adaptive methods to solve free boundary problems of flow through porous media, Int. J. for Num. and Anal. Meth. in Geomech., 11, 17-31.

Comincioli, V., 1974, A comparison of algorithms for some free boundary problems, Publ. Lab. Analisi Numerica CNR Pavia, 79, 38 pp.

Comincioli, V. and Guerri, L., 1976, Numerical solution of free boundary problems in seepage flow with capillary fringe, Comput. Meth. Appl. Mech. Eng., 7, 153-178.

Cryer, C.W., 1976a, A survey of steady state porous flow free boundary problems, MRC Techn. Summary Rep. 1657, Univ. Wisconsin, Madison, 135 pp.

Cryer, C.W., 1976b, A survey of trial free-boundary methods for the numerical solution of free boundary problems, MRC Techn. Summary Rep. 1693, Univ. Wisconsin, Madison, 156 pp.

Cryer, C.W., 1977, A bibliography of free boundary problems, MRC Techn. Summary Rep. 1793, Univ. Wisconsin, Madison.

Desai, C.S., 1976, Finite element residual schemes for unconfined flow, Int. J. for Num. Meth. in Engr., Short Communication, 10, 1415-1418.

Desai, C.S. and Baseghi, B., 1989, Theory and verification of residual flow procedure for 3-D free surface seepage, Advances in Water Resources (in press).

Desai, C.S. and Li, G.C., 1983, A residual flow procedure and application for free surface flow in porous media, Advances in Water Resources, 6, 27-35.

Duvaut, G. and Lions, J.L., 1976, Inequalities in mechanics and physics, Grund. der Math. Wiss., 219, Springer, Berlin.

Friedman, A., 1976a, The shape and smoothness of the free boundary for some elliptic variational inequalities, Ind. Univ. Math. J., 25, 103-118.

Friedman, A., 1976b, A problem in hydraulics with non-monotone free boundary, Ind. Univ. Math. J., 25, 577-592.

Friedman, A. and Jensen, R., 1977, Convexity of the free boundary in the Stefan problem and in the dam problem, Arch. Rat. Mech. Anal., 67, 1-24.

Gilardi, G., 1976, Studio di una famiglia di disequazioni quasivariazionali connessa con un problema di frontiera libera, Boll. U.M.I., 13 – B, 138-159.

Glowinski, R., 1976, Introduction to the approximation of elliptic variational inequalities, No. 76006, Universite Paris VI, Laboratoire Analyse Numerique, L.A., 189.

Glowinski, R., Lions, J.L., and Tremolieres, R., 1976, Analyse Numerique des Inequations Variationnelles, 2 vols., Dunod.

Jensen, R., 1977, Structure of the non-monotone free boundaries in a filtration problem, Ind. Univ. Math. J., 26, 1121-1135.

Kinderlehrer, D., 1976, The free boundary determined by the solution to a differential equation, Ind. Univ. Math. J., 25, 195-208.

Lewy, H. and Stampacchia, G., 1969, On the regularity of the solution of a variational inequality, Comm. Pure Appl. Math., 22, 153-188.

Lewy, H. and Stampacchia, G., 1971, On the existence and smoothness of solutions of some non-coercive variational inequalities, Arch. Rat. Mech. Anal., 41, 241-253.

Lin, C.S., Bruch, J.C., Jr., Sloss, J.M., and Comincioli, V., 1984, On the solution of transient free surface flow problems in porous media by a fixed domain method, J. of Hydrology, 71(3/4), 353-376.

Lions, J.L. and Stampacchia, G., 1967, Variational inequalities, Comm. Pure Appl. Math., 20, 493-519.

Marini, L.D. and Pietra, P., 1986, Fixed-point algorithms for stationary flow in porous media, Comp. Meth. in Appl. Mech. and Engr., 56, 17-45.

Mosco, U., 1973, An introduction to the approximate solution of variational inequalities, Constructive Aspects of Functional Analysis, Edizioni Cremonese, Roma, 499-685.

Oden, J.T. and Kikuchi, N., 1980, Theory of variational inequalities with applications to problems of flow through porous media, Int. J. Engr. Sci., 18(10), 1173-1284.

Pietra, P., 1982, An up-wind finite element method for a filtration problem, R.A.I.R.O. Anal. Num., 16(4), 463-481.

Rank, E. and Werner, H., 1986, An adaptive finite element approach for the free surface seepage problem, Int. J. for Num. Meth. in Engr., 23, 1217-1228.

Remar, J., Bruch, J.C., Jr., and Sloss, J.M., 1984, Numerical solutions to some free surface flows through nonhomogeneous media, Int. J. for Num. Meth. in Engr., 20, 143-167.

Wang, K.-P. and Bruch, J.C., Jr., 1989, Solutions of a steady state free surface seepage problem on a hypercube concurrent computer, Engr. Comp., 6(3), 225-236.

Westbrook, D.R., 1985, Analysis of inequality and residual flow procedures and an iterative scheme for free surface seepage, Int. J. for Num. Meth. in Engr., 21, 1791-1802.

White, R.E., 1987, Multisplittings and parallel iterative methods, Comp. Meth. in Appl. Mech. and Engr., 64, 567-577.

Transport in Porous Media **5**: 651–666, 1991.
© 1991 *Kluwer Academic Publishers. Printed in the Netherlands.*

Qualitative Mathematical Analysis of the Richards Equation

B. H. GILDING
Faculty of Applied Mathematics, University of Twente, P.O. Box 217, 7500 AE Enschede, The Netherlands

(Received: 16 October 1989; in final form: 5 October 1990)

Abstract. The Richards equation is widely used as a model for the flow of water in unsaturated soils. For modelling one-dimensional flow in a homogeneous soil, this equation can be cast in the form of a specific nonlinear partial differential equation with a time derivative and one spatial derivative. This paper is a survey of recent progress in the pure mathematical analysis of this last equation. The emphasis is on the interpretation of the results of the analysis. These are explained in terms of the qualitative behaviour of the flow of water in an unsaturated soil which is described by the Richards equation.

Key words. Mathematical analysis, partial differential equation, nonlinear, diffusion, advection, Fokker–Planck, qualitative behaviour, unsaturated soils, wetting-fronts.

0. Nomenclature

Roman letters

- a coefficient in second-order diffusion term of equation
- b coefficient in first-order advection term of equation
- D soil-moisture diffusivity $[L^2T^{-1}]$
- h pressure head [L]
- H quarter-plane domain for Cauchy-Dirichlet problem $[L] \times [T]$
- K hydraulic conductivity scalar $[LT^{-1}]$
- $\mathbf{K}$ hydraulic conductivity tensor $[LT^{-1}]$
- q soil-moisture flux scalar $[LT^{-1}]$
- $\mathbf{q}$ soil-moisture flux vector $[LT^{-1}]$
- r dummy variable
- R rectangle $[L] \times [T]$
- s dummy variable
- s^* representative value of dummy variable
- S half-plane domain for Cauchy problem $[L] \times [T]$
- t time [T]
- u unknown solution of partial differential equation
- u_0 initial-value function
- v soil-moisture velocity scalar $[LT^{-1}]$
- $\mathbf{v}$ soil-moisture velocity vector $[LT^{-1}]$

x distance, position [L]
z elevation [L]

Greek letters
α angle to horizontal [−]
ζ interface [L]
θ water content [−]
θ_r irreducible water content [−]
θ_s water content at saturation [−]
λ directional parameter [−]
σ test parameter in criterion for finite speed of propagation [LT^{-1}]
σ_0 minimum speed of propagation [LT^{-1}]
ϕ test function
Ω arbitrary domain [L] × [T]

1. Introduction

Neglecting the effects of heat and density gradients, sorption, osmosis and plant abstraction, the flow of water in a soil may be described by Darcy's law

$$\mathbf{q} = -\mathbf{K}\,\mathrm{grad}(h + z) \tag{1}$$

and conservation of mass

$$\frac{\partial \theta}{\partial t} + \mathrm{div}\,\mathbf{q} = 0. \tag{2}$$

Here $\mathbf{q}$ denotes soil-moisture flux, $\mathbf{K}$ is the hydraulic conductivity tensor, h pressure head, z elevation, θ water content, t time and the differential operators 'grad' and 'div' are defined in terms of a fixed coordinate system. The vector $\mathbf{q}$ denotes the bulk flux in the porous medium and is related to the velocity $\mathbf{v}$ of the moisture by

$$\mathbf{q} = \theta \mathbf{v}.$$

Supposing that the soil is isotropic and that the flow is essentially one-dimensional in a direction which is inclined at an angle α to the horizontal, (1) and (2) become

$$q = -K \frac{\partial}{\partial x}(h + \lambda x) \tag{3}$$

and

$$\frac{\partial \theta}{\partial t} = -\frac{\partial}{\partial x} q, \tag{4}$$

respectively. In (3) and (4), q and K are scalars, x denotes distance in the direction of flow, and $\lambda = \sin \alpha$ (see Figure 1). Combining (3) with (4) yields

$$\frac{\partial \theta}{\partial t} = \frac{\partial}{\partial x}\left(K \frac{\partial h}{\partial x} + \lambda K\right). \tag{5}$$

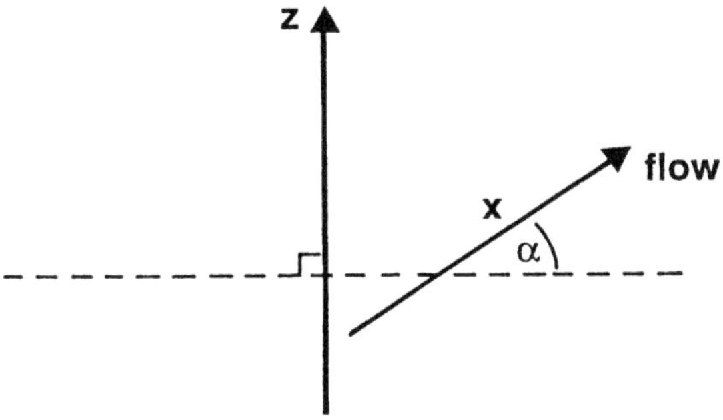

Fig. 1. Direction of flow.

If, furthermore, the soil is supposed to be homogeneous and unsaturated, and hysteresis may be neglected, then K and h may be considered as single-valued functions of θ. In this case, Equation (5) may be rewritten as

$$\frac{\partial\theta}{\partial t} = \frac{\partial}{\partial x}\left(D(\theta)\frac{\partial\theta}{\partial x}\right) + \lambda\frac{\partial}{\partial x}(K(\theta)), \tag{6}$$

where

$$D(\theta) = K(\theta)\frac{\mathrm{d}h}{\mathrm{d}\theta}(\theta)$$

denotes soil-moisture diffusivity [3, 5, 9, 20, 24].

Equation (6) is known as the one-dimensional *Richards equation* formulated in terms of water content [21]. In certain circles [4, 20], this equation is also referred to as the *nonlinear Fokker–Planck equation* because of its resemblance to the celebrated (linear) Fokker–Planck equation occurring in the theory of statistical mechanics [8]. The first-order term on the right-hand side of (6) is frequently called the gravity term. If the flow is horizontal, then $\lambda = 0$, whilst if the flow is vertical in an upwards direction $\lambda = 1$, and if the flow is vertical in a downward direction $\lambda = -1$.

Typical illustrations of the $h(\theta)$ and $K(\theta)$ relationships as found in standard references [3, 9, 20, 24] are shown in Figures 2 and 3. Here θ_s denotes the water content at saturation and θ_r the irreducible water content. The irreducible water content is taken to be zero in the majority of references.

In unsaturated soils, the pressure head is negative and is equated in magnitude with soil-mositure tension or suction. As the water content in a soil progressively increases, it is generally conceived [9, 24] that the smaller pores in which the capillary forces are the greatest are filled before the larger pores. Hence, pressure head h is an increasing function of water content θ. Contemporaneously, since the larger pores in the soil are the most effective conductors, the hydraulic conductivity increases as the water content increases and does so at a progressively greater rate.

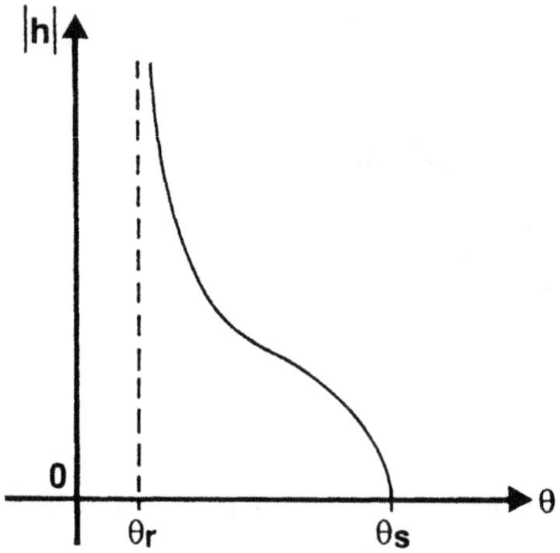

Fig. 2. Typical $h(\theta)$ relationship.

Consequently, hydraulic conductivity K is pictured as an increasing convex function of θ. This function assumes the value zero when the water content assumes its irreducible value.

A typical illustration [9, 24] of the soil-mositure diffusivity relationship including moisture transport in the vapour phase is shown in Figure 4.

The purpose of this paper is to review recent progress [15–17] in the pure mathematical analysis of the equation

$$u_t = (a(u))_{xx} + (b(u))_x \tag{7}$$

in which subscripts denote partial differentiation, and to relate this to the study of Equation (6).

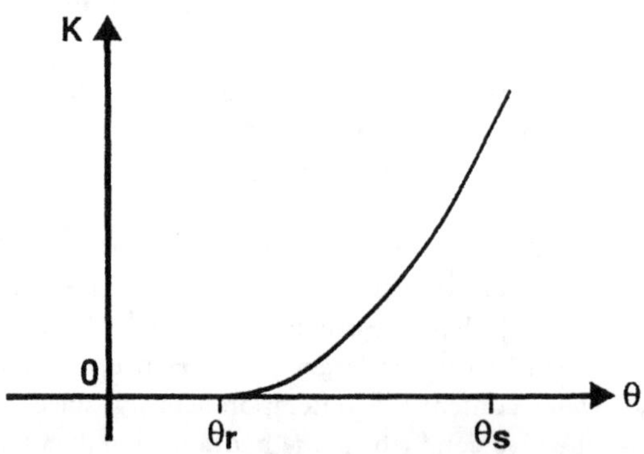

Fig. 3. Typical $K(\theta)$ relationship.

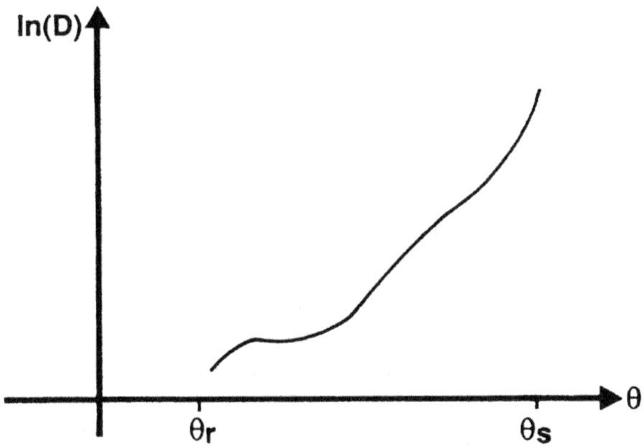

Fig. 4. Typical $D(\theta)$ relationship.

The mathematical analysis of (7) has been predominantly concerned with nonnegative solutions of the equation under the assumptions that the coefficients $a(s)$ and $b(s)$ are defined and continuous for nonnegative arguments s, $a(s)$ and $b(s)$ are twice continuously differentiable for positive arguments s with second derivatives which are what is mathematically known as locally Hölder continuous,

$$a'(s) > 0 \quad \text{for } s > 0 \tag{8}$$

and

$$a(0) = b(0) = 0. \tag{9}$$

Identifying the unknown u in (7) with $\theta - \theta_r$ in (6), a with the indefinite integral of D, and b with λK, Equation (7) can be seen to be equivalent to Equation (6). The hypothesis (8) is equivalent to the positivity of the soil-moisture diffusivity coefficient for water contents above the irreducible level. The hypothesis $a(0) = 0$ in (9) is of no significance since this amounts to no more than a fixing of the constant of integration in the identification of a with the integral of D. The hypothesis $b(0) = 0$ though is equivalent to the datum that the unsaturated hydraulic conductivity tends to zero as the water content approaches its irreducible value.

Of particular note in what follows is that the quantity $-((a(u))_x + b(u))$ can be identified with the flux q in the derivation of the Richards equation and, subsequently, that $-((a(u))_x + b(u))/u$ can be identified with the one-dimensional soil-moisture velocity v for the flow described by this equation.

The survey of the recent mathematical work on Equation (7) will be addressed to the topic of what this says about the acceptability of (6) as a model of the flow of water in unsaturated soils.

2. Solvability

For the soil-physicist or other practitioner considering using the Richards equation (6) for modelling the flow of water in an unsaturated soil, the first question which must spring to mind is: *Given suitable initial and boundary conditions, is there a solution to the equation*? For the mathematician this question is the question of *existence*.

The blunt answer to this question is: No! For many years now it has been known that even with the most regular initial and boundary data it is not possible to find a classical solution to an equation of the type (7) without particularly restrictive conditions on the coefficients a and b which are unrealistic in the present situation [19]. By a classical solution, one means a function $u(x, t)$ for which one can define the partial derivatives $\partial u/\partial t$, $\partial^2(a(u))/\partial x^2$ and $\partial(b(u))/\partial x$ at every point in the domain under consideration in the sense of the limit process with which partial derivatives are defined in any introductory calculus course, and for which one can equate them by (7) at every point in the domain. However, the answer to the question is: Yes; if one considers solutions defined in a suitable abstract mathematical sense.

There are many ways in which the notion of a solution of (7) can be mathematically abstracted [1, 6, 7, 10, 14, 16, 22]. One gambit is to observe that if a function $u(x, t)$ is a classical solution of Equation (7) in some domain Ω, then taking any rectangle $R = (x_1, x_2) \times (t_1, t_2]$ in Ω and any nonnegative function ϕ which is defined and sufficiently smooth in R and which vanishes for $x = x_1$ and $x = x_2$, multiplying (7) by ϕ and integrating by parts one obtains the identity

$$\begin{aligned}\iint_R \{u\phi_t + a(u)\phi_{xx} - b(u)\phi_x\}\, dx\, dt &= \int_{x_1}^{x_2} \{u(x, t_2)\phi(x, t_2) - u(x, t_1)\phi(x, t_1)\}\, dx + \\ &+ \int_{t_1}^{t_2} \{a(u(x_2, t))\phi_x(x_2, t) - a(u(x_1, t))\phi_x(x_1, t)\}\, dt.\end{aligned} \tag{10}$$

A *generalized solution* of Equation (7) in a domain Ω is defined as a function $u(x, t)$ which is continuous, nonnegative, and bounded in Ω and on the boundary of Ω and which satisfies the integral identity (10) for any rectangle R and function ϕ with the afore-stated properties. In this way, any classical solution is a generalized solution. However, the converse is not true.

With the above notion of a generalized solution, various existence theorems for initial and boundary value problems for Equation (7) have been proven [16]. Thus if one considers solutions in this sense, the question of existence has been answered.

The next most obvious question is: *Given suitable initial and boundary conditions, is there only one solution*? For the mathematician this is the question of *uniqueness*.

The answer to this question is also in the affirmative considering the generalized solutions of Equation (7).

For the mathematician the situation is now clear-cut. For fairly respectable problems there is one and only one solution. The natural question that the soil-physicist will then ask though is: *Does the generalized solution have any physical relevance*? To answer this question one may consider the mathematical aspect of *regularity*.

To discuss the regularity of generalized solutions of Equation (7), for simplicity, attention will be restricted to the *Cauchy problem*; this is to say, the problem of solving (7) in the domain

$$S = \{(x, t) : -\infty < x < \infty, 0 < t < \infty\}$$

subject to the initial condition

$$u(x, 0) = u_0(x) \quad \text{for } -\infty < x < \infty,$$

where u_0 is a given function. It is hypothesized that u_0 is nonnegative, continuous and bounded.

Inherent in the definition of a generalized solution $u(x, t)$ of this problem is the property:

- u is bounded and continuous in the domain $\bar{S} = \{(x, t) : -\infty < x < \infty, 0 \leqslant t < \infty\}$.

The following has also been proven [15, 16]:

- $(a(u))_x + b(u)$ is bounded in the mathematical sense of distributions in the set $\{(x, t) : -\infty < x < \infty,\ t_0 \leqslant t < \infty\}$ for every $t_0 > 0$.
- $(a(u))_x + b(u)$ is bounded in the sense of distributions in S if this quantity is initially bounded for $t = 0$ in the sense of distributions.
- $(a(u))_x + b(u)$ is continuous in S and $((a(u))_x + b(u))/u$ is bounded in the sense of distributions in S if the latter quantity is initially bounded in the sense of distributions.
- $(a(u))_x + b(u)$ is continuous in the set $\{(x, t) : -\infty < x < \infty,\ t_0 \leqslant t < \infty\}$ and $((a(u))_x + b(u))/u$ is bounded in the sense of distributions in this set for every $t_0 > 0$ if $a'(s)/s$ is integrable and $b(s)/s$ is bounded in a neighbourhood of $s = 0$. (Strictly speaking this is a simplification of the results. The reader is referred to [15] for details.)
- u is a classical solution of Equation (7) at any point x and time t in S where $u(x, t) > 0$.

An explanation of what is meant by 'in the sense of distributions' may be found for example in [23].

To interpret the above results for the Richards equation, recall that the quantity $-((a(u))_x + b(u))$ is identified with the flux q in the derivation of (6), and that the quantity $-((a(u))_x + b(u))/u$ is analogous to the soil-moisture velocity v. In this light, the mathematical regularity results may be explained as indicating that for the appropriately-defined generalized solutions of (6):

- The water content θ is continuous and bounded.
- The soil-moisture flux q becomes bounded after an infinitestimal time span irrespective of the initial soil-moisture distribution.
- The soil-moisture flux q remains bounded if one begins with an initial soil-moisture distribution with bounded flux.
- The soil-moisture flux q is continuous and the soil-moisture velocity v remains bounded if one begins with an initial soil-moisture distribution with bounded velocity.
- Irrespective of the initial soil-moisture distribution, the soil-moisture flux q becomes continuous and the soil-moisture velocity v becomes bounded after an infinitesimal time-span if the soil-moisture diffusivity $D(\theta)$ and hydraulic conductivity $K(\theta)$ relationships satisfy certain constraints. (The significance of these constraints is a topic which will be returned to later.)
- The equation is solvable in the classical sense after all at any point and instant where the water content is not irreducible.

Thus, albeit that one has to consider solutions of (6) in some abstract mathematical sense, given appropriate initial conditions, these solutions turn out to possess all the properties which one would physically consider necessary, viz. continuous water content, continuous flux and bounded velocity.

Furthermore, since Equation (6) is trivially satisfied by the constant solution $\theta = \theta_r$ inside sub-domains where the water content is irreducible, the only points in which the equation is not solvable in the classical sense and which, therefore, necessitate the introduction of the mathematical notion of a generalized solution, are those which verge on areas where the water content approaches its irreducible value.

With hindsight this is of course wholly consistent with the expounded derivation of (6), since the expressions $h(\theta)$ and $\mathrm{d}(h(\theta))/\mathrm{d}\theta$ only have a meaning if the water content is above its irreducible level.

3. The Phenomenon of Wetting-Fronts

It should be evident from the previous section that the crucial difficulty with the mathematics of Equation (6) is occasioned by the possible occurrence of points demarcating regions in which the water content is above its irreducible level. Since in those areas where the water content is irreducible the medium is to all intents and purposes dry, such a point is equivalent to the physical phenomenon of a *wetting-front*. The question to be asked is then: *Do wetting-fronts really occur*? This question can be epitomized by supposing that one takes a dry soil and introduces water into a restricted area. One would then like to know whether at any later time the water still occupies a restricted area in an otherwise dry medium, or whether the distinction between wet and dry regions immediately disappears. (For an illustration in a situation where the irreducible water content $\theta_r = 0$, see Figure 5.)

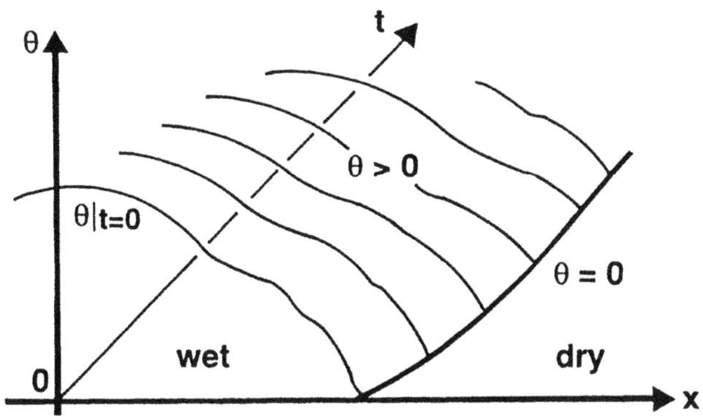

Fig. 5. Visualization of a wetting-front.

The latter question can be translated as follows. Consider the solution $u(x, t)$ of the Cauchy problem for Equation (7) and define

$$\zeta(t) = \sup\{x : -\infty < x < \infty, u(x, t) > 0\} \tag{11}$$

for all $t \geqslant 0$. This is to say, let $\zeta(t)$ denote the least upper bound of the points where the solution is positive at any time t. Then if

$$-\infty < \zeta(0) < \infty, \tag{12}$$

is it true that

$$-\infty < \zeta(t) < \infty \quad \text{for any } t > 0?$$

If this is the case one speaks of *finite speed of propagation* in mathematical terms and refers to the point $\zeta(t)$ which corresponds to a wetting-front as an *interface*. The expression 'finite speed of propagation' may be declared in terms of the Richards' equation by observing that if there are no wetting-fronts in the sketched situation, then water must have instantaneously penetrated arbitrarily far into the medium and, thus, have progagated with an infinite speed.

It has been proven [15, 18] that in the above-mentioned context Equation (7) displays finite speed of propagation if and only if

$$a'(s)/\max\{s, b(s)\} \text{ is integrable from 0 to } s^* \tag{13}$$

and there is a constant σ such that

$$-b(s) < \sigma s \quad \text{for } 0 < s \leqslant s^* \tag{14}$$

for some $s^* > 0$.

This result has been obtained with the aid of two mathematical tools which have been established for generalized solutions of (7). The first of these is the *comparison principle*. This says that given two generalized solutions of Equation (7), u_1 and u_2

say, in some domain Ω of the form $(x_1, x_2) \times (t_1, t_2]$, if $u_1 \geqslant u_2$ at every point on the boundary of Ω excluding the line $t = t_2$ then $u_1 \geqslant u_2$ in the whole of Ω. The other tool is the *integral identity*

$$\int_{-\infty}^{\infty} u(x, t)\, \mathrm{d}x = \int_{-\infty}^{\infty} u_0(x)\, \mathrm{d}x \quad \text{for any } t > 0 \tag{15}$$

for generalized solutions $u(x, t)$ of the Cauchy problem for Equation (7). Both of these tools may be couched in physical terms.

The comparison principle basically says that given two identical soil samples, if the first is initially wetter than the other and is continually maintained wetter on the perimeter than the other, then at any later time this sample will also be wetter at every point in the interior. The argument may be refined by the observation that if one also injects water into the first sample or abstracts it from the second, the conclusion remains valid. In mathematical terms this is tantamount to the comparison principle for generalized *supersolutions* and *subsolutions*.

Noting the analogy between u and θ and recollecting that in the derivation of Equation (6) it has implicitly been assumed that the water density is constant, the integral identity (15) is no more than a statement of *conservation of mass* for the soil water.

Returning to the question of the existence of wetting-fronts, recall that u in (7) is equivalent to $\theta - \theta_r$ in (6), the derivative a' is analogous to the diffusivity coefficient $D(\theta)$, and b is equivalent to $\lambda K(\theta)$. Whilst, $K(\theta) > 0$ for $\theta > \theta_r$, and $\lambda = 1$ if the flow is directed upwards, $\lambda = 0$ if the flow is horizontal, and $\lambda = -1$ if the flow is directed downwards. In this light, the somewhat esoteric condition (13), (14) infers that in one-dimensional flow wetting-fronts can occur in an upward direction if and only if

$$D(\theta)/\max\{\theta - \theta_r, K(\theta)\} \text{ is integrable from } \theta_r \text{ to } \theta_s. \tag{16}$$

Whilst wetting-fronts can occur in a horizontal direction if and only if

$$D(\theta)/(\theta - \theta_r) \text{ is integrable from } \theta_r \text{ to } \theta_s. \tag{17}$$

Finally, wetting fronts can occur in a downward direction if and only if (17) holds and $K(\theta)$ lies under a straight-line when visualized as in Figure 6. Note that in this case, (16) and (17) are equivalent.

Consequently if one believes in wetting-fronts in all directions the soil-moisture diffusivity relationship must satisfy (17) and the $K(\theta)$ relationship must appear as in Figure 6. Observe though that if the $K(\theta)$ curve is convex, which experiment and argument have indicated is the case, the latter criterion is automatically satisfied.

It is interesting to note that the last-mentioned constraints on the relations $D(\theta)$ and $K(\theta)$ are precisely those under which the optimal regularity results discussed earlier were obtained. Thus, under these constraints, not only can one conclude that wetting-fronts are admissible but also one can conclude that, in any given soil profile, the soil-moisture flux will be continuous and the soil-moisture velocity will

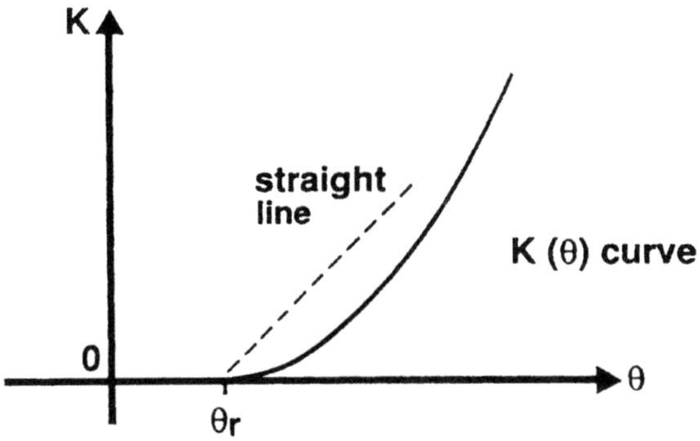

Fig. 6. Condition on $K(\theta)$ for existence of downward wetting-front.

be bounded after an infinitesimal time irrespective of how irregular the initial soil-moisture distribution may be. If the initial soil-moisture velocity is bounded, then the flux will be continuous and the velocity will be bounded from the start.

4. Properties of Wetting-Fronts

Continuing the analysis of the occurrence of wetting-fronts, a number of physically relevant properties of the interface $\zeta(t)$ defined by (11) for the solution $u(x, t)$ of the Cauchy problem for Equation (7) have been proven under conditions (12)–(14).

The first of these properties is that there is a *minimum speed of propagation*. To be specific, it has been proven [15] that for any $t_1 \geqslant 0$ and $t_2 \geqslant t_1$ there holds

$$\zeta(t_2) \geqslant \zeta(t_1) + \sigma_0(t_2 - t_1),$$

where

$$\sigma_0 = \limsup_{s \to 0} - b(s)/s. \tag{18}$$

In terms of Equation (6), this minimum speed

$$\sigma_0 = \limsup_{\theta \to \theta_r} - \lambda \frac{K(\theta)}{\theta - \theta_r} = -\lambda K'(\theta_r).$$

assuming that the latter exists. Thus, when $\lambda = -1$ or when $\lambda = 0$, by the nonnegativity of the hydraulic conductivity function, one has $\sigma_0 \geqslant 0$. Whereas when $\lambda = 1$, the minimum speed $\sigma_0 < 0$ if $K'(\theta_r) > 0$, and $\sigma_0 = 0$ if $K'(\theta_r) = 0$.

This means automatically that in a vertical profile any wetting-front demarcating the lower boundary of a wetted region cannot move upwards, whilst in a horizontal profile any wetting-front must propagate outwards. Furthermore, any upper wetting-front demarcating a wetted region in a vertical profile cannot move downwards if one can say that $K'(\theta_r) = 0$. See Figure 7. Subsequently, since the evidence indicates that $K(\theta)$ relationships have a convex form such as that shown in Figures

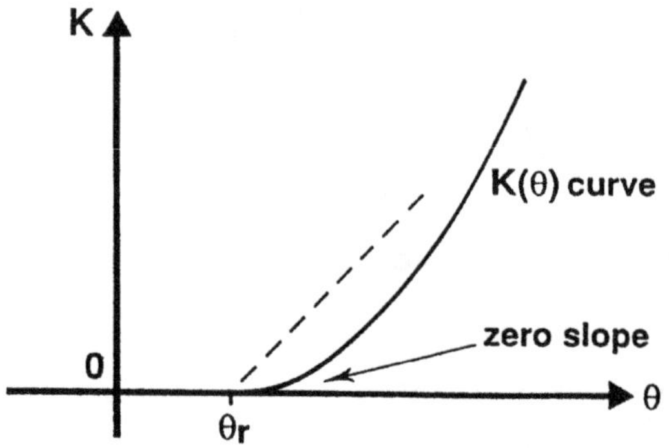

Fig. 7. Condition on $K(\theta)$ for non-contraction of wetting-front.

3, 6 and 7, whatever the direction, wetted regions cannot shrink. The name wetting-front is consequentially not misplaced.

The above answers the question: *Is soil-moisture retained*? A consequence of the observation that a wetted region can only expand is that at any given place once the water content has increased above its irreducible level it can never again drop back to this level. Consequently, there is always some *retention*, albeit in real terms perhaps in increasingly smaller concentrations in increasingly narrower pores.

The next question which can be answered by the mathematical theory of (7) is: *Does the movement of a wetting-front exhibit jumps*? Casting this question in the negatory form for Equation (7) it becomes: Is the interface $\zeta(t)$ continuous? The answer here is: Yes, if the σ_0 defined by (18) is not infinite in magnitude [15]. Whence, under the previously-mentioned criterion of zero slope of the $K(\theta)$ relationship at the irreducible water level ($\sigma_0 = 0$) a wetting-front can indeed only move continuously, i.e. without jumps.

(Whatever the value of σ_0, it is known that the interface $\zeta(t)$ is lower semi-continuous and continuous from the right. However, if $\sigma_0 = -\infty$, the upper semi-continuity of the interface from the left is still an open mathematical question [15].)

A further property to be considered of the interface $\zeta(t)$ relates to the occurrence of a wetting-front as a material surface. Physically, one would expect that the advance of a wetting-front is determined by the speed at which the water particles forming the front travel. The question to be posed then is: *Does a wetting-front move at a speed equal to the velocity of the water at the front*?

In terms of the solution of the Cauchy problem for Equation (7) this question can be reformulated as: Is

$$\zeta'(t) = \lim_{x \uparrow \zeta(t)} [-((a(u))_x + b(u))/u](x, t)$$

for any $t \geqslant 0$? The answer to this question is: Yes, in some limited sense. For details, see [15, 25]. However, as yet with the exception of the case when $b \equiv 0$ and

a takes on an idealized form [2, 19], here the mathematical theory is far from being complete.

5. Penetration

The final question to be addressed in this survey is that of the penetrability of water in a soil. Supposing that one introduces moisture into a restricted region of a dry soil, one may ask: How far will the moisture eventually infiltrate, or, *Is moisture penetration boundless*?

To answer this last question, an alternative mathematical problem to that previously considered will be examined. This is the *Cauchy–Dirichlet problem* of solving Equation (7) in the domain

$$H = \{(x, t) : 0 < x < \infty, 0 < t < \infty\}$$

subject to the initial condition

$$u(x, 0) = u_0(x) \quad \text{for } 0 \leqslant x < \infty$$

and the boundary condition

$$u(0, t) = 0 \quad \text{for } 0 < t < \infty. \tag{19}$$

Here $u_0(x)$ is hypothesized to be a given bounded nonnegative continuous function for $0 \leqslant x < \infty$ which is compatible with the boundary condition (19) via the constraint $u_0(0) = 0$. This problem admits a unique generalized solution, $u(x, t)$, and defining

$$\zeta(t) = \sup\{x : 0 \leqslant x < \infty, \quad u(x, t) > 0\},$$

when $\zeta(0) < \infty$ and (13) and (14) hold, it is known that $\zeta(t) < \infty$ for all $t > 0$ [15–17]. Since plainly if u_0 is identically zero, the solution to this problem is $u \equiv 0$, it will be supposed that $\zeta(0) > 0$.

With regard to the original question, the above problem may be viewed as the worst possible case. For, in essence, it embodies a semi-infinite profile in which the soil is *desiccated*, i.e. $\theta = \theta_r$, at the only accessible boundary.

For the new problem the question is: Does

$$\zeta(t) \to \infty \text{ as } t \to \infty? \tag{20}$$

The following has been proven [17]:

(i) If

$$b(s) \leqslant 0 \quad \text{for } 0 < s \leqslant s^* \tag{21}$$

for some $s^* > 0$, then (20) holds.

(ii) If

$$a'(s)/s \text{ is integrable from } 0 \text{ to } s^*,$$

$$a'(s)/(\textstyle\int_0^s a'(r)/r \, dr) \text{ is bounded for } 0 < s \leqslant s^*$$

and

$$b'(s)/(\textstyle\int_0^s a'(r)/r \, dr) \text{ is bounded for } 0 < s \leqslant s^*$$

for some $s^* > 0$, then (20) holds.

(iii) If

$$b(s) > 0 \quad \text{for } 0 < s \leqslant s^*$$

and

$$a'(s)/b(s) \text{ is integrable from } 0 \text{ to } s^* \tag{22}$$

for some $s^* > 0$, then (20) does not hold.

Otherwise, when the parameter σ_0 defined by (18) is zero, the question is open.

The tools with which these results have been proven are the comparison principle for generalized solutions of (7) mentioned earlier and the integral identity for solutions of the Cauchy–Dirichlet problem

$$\int_0^\infty x u(x, t) \, dx = \int_0^\infty x u_0(x) \, dx - \int_0^t \int_0^\infty b(u(x, \tau)) \, dx \, d\tau$$

for any $t > 0$. The latter may be construed as a statement of *conservation of momentum* for the corresponding flow described by the Richard's equation.

Since b in (7) is identified with $\lambda K(\theta)$ in (6) and K is always nonnegative, the condition (21) is automatically fulfilled by Equation (6) when $\lambda = -1$ and when $\lambda = 0$. Thus, the result (i) above implies that irrespective of the soil characteristics, for soil-moisture infiltration in the direction of gravitational pull or in a horizontal direction, the penetration of water cannot be limited. In other words, under circumstances in which soil-moisture infiltration can be regarded as a one-dimensional phenomenon, once moisture is introduced into a vertical profile lower-lying regions will without fail become wetted at some later time. Similarly, once water is introduced into a horizontal soil profile, every point will always eventually become wetted.

The second result (ii) above indicates that the penetration of moisture also cannot be limited in an upward direction when the soil-moisture diffusitivity and hydraulic conductivity relations fulfil the implied constraints. The constraints themselves are not at all preposterous. They are, for instance, satisfied by the closed-form $D(\theta)$ and $K(\theta)$ relationships discussed in [12, 13]. By way of speaking, in soils whose characteristics conform to these constraints, capillary suction is a sufficiently powerful mechanism with regard to upward water penetration to override the influence of downward gravitational pull.

The final result (iii) above is especially curious with regard to the Richards equation. When $\lambda = 1$, a'/b corresponds exactly with

$$D(\theta)/K(\theta) = \frac{\mathrm{d}h}{\mathrm{d}\theta}(\theta)$$

in (6). So, fulfilment of condition (22) means that $h(\theta)$ must tend to some bounded limit as θ tends to θ_r. The latter though appears to be contradicted by the $h(\theta)$ relationships to be found in the literature [3, 9, 12, 13, 20, 24] where h disappears from the picture at low water contents. Cf. Figure 2.

Should pressure head h remain bounded as water content θ approaches its critical level θ_r, then the result (iii) above infers that soil-moisture penetration in an upward direction may be limited under the application of appropriate boundary conditions. However, since the generally exhibited behaviour of the $h(\theta)$ relationship does not conform to this picture, the answer to the question 'Do wetting-fronts expand without bound' is as yet unresolved for the upward moving wetting-fronts in a vertical profile whose characteristics are not covered by (ii).

6. Discussion and Conclusions

Nowadays, the use of the Richards equation as a model for the flow of water in an unsaturated soil is commonplace [4, 11–13]. Dependent on the application involved, it may appear in the form considered here or formulated in terms of pressure head or suction, for a homogeneous or heterogeneous soil, and with or without plant uptake.

The essential difficulty with the mathematical analysis of the Richards equation is the possible occurrence of wetting-fronts where the equation may not be solvable in the classical sense. The actuality of this possibility is determined by the behaviour of the $K(\theta)$ and $h(\theta)$ relationships as the water content approaches its irreducible level. It is of course questionable how relevant the Richards equation is for describing soil-moisture flow when the water content is extremely low. Among the factors neglected in the derivation of Equation (6) is the movement of water in vapour form which becomes influential at low moisture contents. Perhaps more importantly, the validity of the continuum assumption is open to question at this level. A discussion of various aspects of the justifiability of the Richards equation as a model for the flow of water in an unsaturated soil from the view-point of a soil-physicist can be found in [20]. One point which should be highlighted is that of the (un)feasibility of performing experiments to determine $K(\theta)$ and $h(\theta)$ relationships at arbitrary low water contents.

Notwithstanding the above remarks, the mathematical analysis surveyed in this paper has shown that in the form considered here the Richards equation can be solved in a meaningful way. Moreover, it has led to the identification of criteria on the $K(\theta)$ and $h(\theta)$ relationships which characterize the occurrence and the behaviour of wetting-fronts.

The kind of analysis surveyed in this paper does not only serve to justify the use of such an equation as a model. In practice, experience has repeatedly taught that an understanding of the mathematical properties of an equation is an essential key to the development of efficient computational schemes for the numerical solution of the equation in hand.

References

1. Alt, H. W. and Luckhaus S., 1983, Quasilinear elliptic-parabolic differential equations, *Math. Z.* **183**, 311–341.
2. Aronson, D. G., 1986, The porous medium equation, in A. Fasano and M. Primicerio (eds.), *Nonlinear Diffusion Problems*, Lecture Notes in Mathematics 1224, Springer-Verlag, Berlin, pp. 1–46.
3. Bear, J., 1972, *Dynamics of Fluids in Porous Media*, American Elsevier, New York.
4. Bear, J., 1979, *Hydraulics of Groundwater*, McGraw-Hill, New York.
5. Bear, J., Zaslavsky, D., and Irmay, S., 1968, *Physical Principles of Water Percolation and Seepage*, United Nations Educational Scientific and Cultural Organization, Paris.
6. Bénilan, P. and Touré, H., 1984, Sur l'équation générale $u_t = \varphi(u)_{xx} - \psi(u)_x + v$, *C. R. Séances Acad. Sci. Série I Math.* **299**, 919–922.
7. Cannon, J. R. and Mohamed, F. A., 1989, A multifree boundary problem arising in the theory of liquid flow in a porous medium, *Unione Mat. Ital. Boll. B Serie VII* **3**, 69–93.
8. Chandrasekhar, S., 1943, Stochastic problems in physics and astronomy, *Rev. Modern Phys.* **15**, 1–89.
9. Childs, E. C., 1969, *An Introduction to the Physical Basis of Soil Water Phenomena*, Wiley, London.
10. Di Benedetto, E. and Showalter, R. E., 1981, Implicit degenerate evolution equations and applications, *SIAM J. Math. Anal.* **12**, 731–751.
11. Feddes, R. A., Kowalik, P. J., and Zaradny, H., 1978, *Simulation of Field Water Use and Crop Yield*, Centre for Agricultural Publishing and Documentation, Wageningen, The Netherlands.
12. Genuchten, M. T. van, 1980, A closed-form equation for predicting the hydraulic conductivity of unsaturated soils, *Soil Sci. Soc. Amer. J.* **44**, 892–898.
13. Genuchten, M. T. van and Nielsen, D. R., 1985, On describing and predicting the hydraulic properties of unsaturated soils, *Ann. Geophys.* **3**, 615–624.
14. Gilding, B. H., 1977, A nonlinear degenerate parabolic equation, *Ann. Scuola Norm. Sup. Pisa Cl. Sci. 4* **4**, 393–432.
15. Gilding, B. H., 1988, The occurrence of interfaces in nonlinear diffusion-advection processes, *Arch. Rational Mech. Anal.* **100**, 243–263.
16. Gilding, B. H., 1989, Improved theory for a nonlinear degenerate parabolic equation, *Ann. Scuola Norm. Sup. Pisa Cl. Sci. 4* **16**, 165–224.
17. Gilding, B. H., 1989, Localization of solutions of a nonlinear Fokker–Planck equation with Dirichlet boundary conditions, *Nonlinear Anal. Theory Meth. Appl.* **13**, 1215–1240.
18. Gilding, B. H. and Kersner, R., 1992, The characterization of reaction-convection-diffusion processes by travelling waves, in preparation.
19. Kalashnikov, A. S., 1987, Some problems of the qualitative theory of non-linear degenerate second-order parabolic equations, *Russian Math. Surveys* **42**, 169–222. Translation of *Uspekhi Mat. Nauk* **42**, 135–176.
20. Philip, J. R., 1969, Theory of infiltration, in Ven Te Chow (ed.) *Advances in Hydroscience Volume* 5, Academic Press, New York, pp. 215–296.
21. Richards, L. A., 1931, Capillary conduction of liquids through porous mediums, *Physics* **1**, 318–333.
22. Sacks, P. E., 1983, Continuity of solutions of a singular parabolic equation, *Nonlinear Anal. Theory Meth. Appl.* **7**, 387–409.
23. Smoller, J., 1983, *Shock Waves and Reaction-Diffusion Equations*, Springer-Verlag, New York.
24. Swartzendruber, D., 1969, The flow of water in unsaturated soils, in R. J. M. de Wiest (ed.), *Flow through Porous Media*, Academic Press, New York, pp. 215–292.
25. Wang, S., 1983, The lateral boundary of the support of weak solutions of the filtration equation in one dimension, *Acta Math. Sinica* (*Beijing*) **26**, 199–219.

Transport in Porous Media **6**: 667–702, 1991
© 1991 *Kluwer Academic Publishers. Printed in the Netherlands.*

MODELING OF *IN-SITU* BIORESTORATION OF ORGANIC COMPOUNDS IN GROUNDWATER

Chen Y. Chiang

Shell Development Company, Houston, Texas U.S.A.

Clint N. Dawson

Department of Mathematics, University of Chicago, Illinois U.S.A.

Mary F. Wheeler

Department of Mathematics, University of Houston, Texas U.S.A.

Abstract: A convergent numerical method for modeling *in situ* biorestoration of contaminated groundwater is outlined. This method treats systems of transport-biodegradation equations by operator splitting in time. Transport is approximated by a finite element modified method of characteristics. The biodegradation terms are split from the transport terms and treated as a system of ordinary differential equations. Numerical results for vertical cross-sectional flow are presented. The effects of variable hydraulic conductivity and variable linear adsorption are studied.

Keywords: Modified method of characteristics, time-splitting, *in situ* biorestoration, adsorption, contaminant transport.

1 Introduction

In recent years, *in situ* biorestoration has emerged as the most promising technology for the treatment of contaminated groundwater. Numerous field and laboratory studies have indicated that indigenous microorganisms degrade harmful compounds in both aerobic and anaerobic aquifers, and control contaminant movement; see [14, 23, 27, 28] and the references therein. *In situ* biorestoration involves the enhancement of these natural processes by introducing dissolved oxygen and/or other nutrients into the system. As discussed in a recent Environmental Protection Agency (EPA) report [27], *in situ* biorestoration is effective for several reasons:

1) it is environmentally sound, generally producing no toxic byproducts,

2) it typically results in complete degradation of contaminants,

3) it utilizes indigenous microflora and does not introduce potentially harmful organisms,

4) it is fast, safe, and generally economical,

and

5) the treatment moves with the groundwater.

An example of the advantage of this process over standard technology is the Ambler, Pennsylvania field site [22]. Here, a gasoline spill was remediated in 10 months by biorestoration, whereas standard "pump and treat" technology was estimated to take 100 years.

The biorestoration process is physically and chemically complex, involving transport and interaction of substrate, dissolved oxygen, and microorganisms. Many factors must be considered, such as the characteristics of the organisms, the primary substrate and nutrients, cometabolic substrates, contaminant concentration and location, available electron acceptors, and environmental conditions, including sorption, ion exchange, the effect of colloidal particles, soil pH and mineral composition, temperature, rock heterogeneity, and fractures.

Much of the field and laboratory data used to estimate these effects is incomplete and/or inexact, and is scale-dependent.

Presently, a major research emphasis is in laboratory column and batch reactor experiments and intermediate scale laboratory studies [28] to understand and quantify the sorption and biodegradation kinetics. Because in-depth studies on a field scale are expensive, mathematical modeling serves as a valuable link between the experimental studies and scientific theory. It is especially useful in 1) understanding the mechanisms of interactive transport, sorption, and biodegradation, 2) analyzing the sensitivity and significance of various parameters in a model and providing insight into field data collection activities, and 3) efficiently investigating various strategies for remediation and mitigation.

The governing equations describing these processes are coupled nonlinear partial differential equations involving transport, biodegradation, and sorption of substrates, transport and utilization of dissolved oxygen and other nutrients, and the metabolic dynamics and possible transport of microorganisms. In the next section, we discuss some of the specific mathematical models which have been proposed by Molz, Widdowson, and Benefield [20], Borden and Bedient [1], and Kindred and Celia [14]. In Section 3, we describe a numerical method for approximately solving the governing equations arising from the latter two formulations. In our algorithm, an accurate mixed finite element method is used to calculate the flow field. A time-splitting procedure is then employed to solve the transport-biodegradation equations. Transport is approximated by a modified method of characteristics, and the reactions are treated as a system of ordinary differential equations, which are solved by integration in time. Finally, in Section 4, we present numerical results for vertical cross-sectional flow. In these studies, we are particularly interested in the effects of spatially varying hydraulic conductivity, and spatially varying organic carbon content, which influences adsorption.

2 Mathematical models

Various mathematical models have been proposed for describing the biodegradation process. In this section we briefly discuss those proposed by Borden and Bedient, Molz, *et al*, and Kindred and Celia. The latter two models generalize the first; each model includes the biodegradation terms as sources and sinks in the transport equations.

Borden and Bedient, and Molz, *et al*, have formulated equations for aerobic systems with two nutrients (substrate and dissolved oxygen). Borden and Bedient based their model on the Monod kinetics. This formulation describes degradation of a single organic compound and dissolved oxygen utilization by microbes. Molz, *et al*, incorporated a biofilm model of Rittman and McCarty [24] which describes the growth, uptake, and degradation of contaminants within a consortium of microbes which are attached to the soil particles. This approach consists of five coupled equations describing substrate transport, dissolved oxygen transport, utilization of substrate and dissolved oxygen by microcolonies, and microcolony evolution, and requires the knowledge of nineteen input parameters. By assuming that substrate and dissolved oxygen utilization are constant multiples of substrate and dissolved oxygen in solution, this model reduces essentially to that of Borden and Bedient.

An extension of these models to include anaerobic biodegradation has been developed by Kindred and Celia [14]. They also generalize the aerobic metabolism with uptake to multiple organic energy substrates, and include cometabolism of nongrowth substrates by aerobic bacteria. The Michaelis-Menten model of uptake kinetics is assumed, and uptake inhibition is included.

All of these models treat sorption as a linear equilibrium isotherm and assume no interaction between sorption and biodegradation.

Numerical simulations using these models have been implemented by the above mentioned researchers. Borden and Bedient [1] and Rifai [23] have modified the Konikow-Bredeheoft [15] code to simulate aerobic biodegradation by assuming an instantaneous stoichiometric biochemical oxidation of the soluble organics. Widdowson, *et al* [32], employ a first-order, finite difference Eulerian-Lagrangian scheme for transport, and use a Newton-Raphson method to treat the nonlinear reaction terms. Their transport scheme is equivalent to a finite difference modified method of characteristics, as analyzed by Douglas and Russell [10].

The one space dimensional numerical experiments by Celia, *et al* [2], utilize an "optimal test function" approach which is closely related to the $\mathcal{H}^{-1}$ Galerkin method [11]. Biodegradation is treated by quadratic extrapolation in time.

Of the above, Borden and Bedient, Rifai, and Widdowson, *et al*, have compared their numerical results with field data. In all of the above mentioned simulations, a constant flow field was assumed.

We would also like to mention more recent numerical work in this area by MacQuarrie, *et al* [16], and MacQuarrie and Sudicky [17].

Another numerical method for approximating the Borden and Bedient equations has been formulated by Wheeler and Dawson [30]. This work represents extensions of earlier work of Douglas and Russell [10], Russell [25], Russell, *et al* [26], and Chiang, *et al* [4] to transport-biodegradation systems. Numerical experiments utilizing this method have been reported in [6, 7, 8, 30, 31]. In the next section, we discuss this method in the context of more general biodegradation models.

3 Numerical Method

In this section, we describe a numerical procedure for solving a general transport-biodegradation model, assuming linear instantaneous adsorption of substrate. This procedure is applicable to the Borden and Bedient model and the various models proposed by Kindred and Celia. We do not discuss application to the formulation of Molz, *et al*, however, time-splitting may be employed in this context as well.

We denote by S_i, $i = 1, \ldots, m_s$, the concentrations of various substrates in solution; O represents the concentration of dissolved oxygen; N_i, $i = 1, \ldots, m_n$, denotes the concentrations of various other nutrients and electron acceptors (such as nitrate and assimilatory nitrogen), and B_i, $i = 1, \ldots, m_b$, the concentrations of various bacterial species.

Let $\mathcal{S} = (S_1, \ldots, S_{m_s})$, $\mathcal{N} = (N_1, \ldots, N_{m_n})$, and $\mathcal{B} = (B_1, \ldots, B_{m_b})$. The equations describing biodegradation of substrates, utilization of dissolved oxygen and nutrients, and microbial metabolism can be written in general form as

$$R_{S_i} \frac{dS_i}{dt} = RS_i(\mathcal{S}, O, \mathcal{N}, \mathcal{B}), \quad i = 1, \ldots, m_s, \tag{1}$$

$$\frac{dO}{dt} = RO(\mathcal{S}, O, \mathcal{N}, \mathcal{B}), \tag{2}$$

$$\frac{dN_i}{dt} = RN_i(\mathcal{S}, O, \mathcal{N}, \mathcal{B}), \quad i = 1, \ldots, m_n, \tag{3}$$

and

$$\frac{dB_i}{dt} = RB_i(\mathcal{S}, \mathcal{O}, \mathcal{N}, \mathcal{B}), \quad i = 1, \ldots, m_b, \tag{4}$$

where R_{S_i} is the retardation factor for substrate i due to adsorption. We note that, in certain cases, the right hand side terms may not explicitly depend on all of the variables.

Combining (1)-(3) with linear transport and instantaneous adsorption we obtain the following system of equations,

$$\phi R_{S_i} \frac{\partial S_i}{\partial t} - \nabla \cdot (D \nabla S_i - \mathbf{u} S_i) = \phi RS_i, \quad i = 1, \ldots, m_s, \tag{5}$$

$$\phi \frac{\partial O}{\partial t} - \nabla \cdot (D \nabla O - \mathbf{u} O) = \phi RO, \tag{6}$$

$$\phi \frac{\partial N_i}{\partial t} - \nabla \cdot (D \nabla N_i - \mathbf{u} N_i) = \phi RN_i, \quad i = 1, \ldots, m_n, \tag{7}$$

$$\mathbf{u} = -K \nabla h, \tag{8}$$

and

$$\nabla \cdot \mathbf{u} = 0, \tag{9}$$

where $D = D(\mathbf{u})$ is the diffusion/dispersion tensor, which includes the effects of molecular diffusion and longitudinal and transverse dispersion, $\mathbf{u} = (u_1, u_2)^T$ = Darcy velocity, h = hydraulic head, K = hydraulic conductivity, and ϕ = porosity. Here we have neglected hydraulic source terms (wells) which may also be included in the system of equations. Combining (5)-(9) and (4) with appropriate initial and boundary conditions completes the mathematical description of the problem. Note that we have assumed microbe transport is negligible. If this is not the case, we obtain equations of the form (5) for those species which are transported.

Solving these equations numerically involves the treatment of sharp fronts, heterogeneities, physical dispersion, and time scale effects. In particular, biodegradation generally needs to be approximated on a much smaller time scale than advection and dispersion. All of these difficulties must be handled in a robust and computationally efficient fashion. We feel that the approach described below satisfies these requirements. The algorithm we have implemented is as follows:

(1) Accurate Darcy velocities and head are calculated using a higher-order mixed finite element procedure.

Steps (2) and (3) are repeated N times where $N\Delta t$ = final time:

(2) A finite element modified method of characteristics is used to treat each of the transport equations, i.e., (5)-(7) with right hand sides set to zero.

(3) The biodegradation equations (1)-(4) are treated as a system of ordinary differential equations, where a second-order, explicit Runge-Kutta method is employed, using many small time steps.

In this paper, we consider only two space dimensions; however, the methodology is easily extendable to three space dimensions.

The approximating spaces we use in step 1 are the Raviart-Thomas spaces with $r = 1$. That is, the hydraulic head approximation is a discontinuous, piecewise bilinear function on a tensor product, rectangular grid. The x-component of velocity (y-component) is approximated by a continuous, piecewise quadratic function in $x(y)$ and a discontinuous, piecewise linear function in $y(x)$. The dimension of these spaces is approximately $12N_xN_y$ where N_x and N_y are the number of intervals in the x and y directions respectively; for more details, see [9]. These approximating spaces for hydraulic head and velocity yield higher order convergence rates than the standard cell-centered finite differences (i.e., Raviart-Thomas approximating spaces with $r = 0$), and do not introduce any grid orientation effects. Moreover, in the mixed finite element method, hydraulic conductivity is treated in an integral formulation, which allows for more accurate modeling of highly varying conductivities. A complete analysis of the mixed formulation and superconvergence results can be found in [21].

The time-splitting algorithm outlined in steps 2 and 3 for solving the transport-biodegradation equations (5)-(7) is most conveniently described by considering a single component, which we denote as C. Let R_C be the retardation factor for component C, with the convention that for non-substrates, $R_C = 1$, and let $\psi = \phi R_C$. Let $\Delta t > 0$, and $t^n = n\Delta t$, for some integer $n \geq 0$. Assume $C^n \equiv C(t^n)$ is known, and $C^{n+1} \equiv C(t^n + \Delta t)$ is desired. Thus, we want to solve

$$\psi\frac{\partial C}{\partial t} - \nabla \cdot D\nabla C + \mathbf{u} \cdot \nabla C = \phi RC \tag{10}$$

for $t^n < t \leq t^{n+1}$ with initial condition C^n. We time-split (10), approximating biodegradation and transport sequentially; denote the biodegradation solution by C^* and the transport solution by $\bar{C}$. First, we solve the system of ordinary differential equations for biodegradation given by (1)-(4), which has components of the form

$$R_C\frac{dC^*}{dt} = RC(\mathcal{S}, \mathcal{O}, \mathcal{N}, \mathcal{B}). \tag{11}$$

This system is solved for $t^n < t \leq t^{n+1}$ with component initial conditions $C^*(t^n) = C^n$. Note that since microorganisms are not transported, this completes the step for these components. Next, transport of substrate, dissolved oxygen, and other nutrients is modeled by solving equations of the form

$$\begin{aligned} \psi\frac{\partial\bar{C}}{\partial t} - \nabla\cdot(D\nabla\bar{C} - \mathbf{u}\bar{C}) &= 0, \qquad t^n < t \leq t^{n+1}, \\ \bar{C}(t^n) &= C^*(t^{n+1}). \end{aligned} \tag{12}$$

This two step procedure gives $\bar{C}(t^{n+1})$ which is an approximation to C^{n+1}.

We approximate C^n for each n in the finite element space of continuous, piecewise bilinears on a rectangular mesh, which we denote as $\mathcal{M}_h$. The system of ordinary differential equations corresponding to (11) is solved at each node in the finite element mesh, using an explicit, second-order Runge-Kutta method with time step Δt_s, which is generally much smaller than the transport time step, depending on the stiffness of the system (11). These solutions are interpolated to obtain an approximation to $C^*(t^{n+1})$ in $\mathcal{M}_h$, denoted by $C_h^*(t^{n+1})$.

The transport solutions are approximated using a finite element modified method of characteristics (MMOC). Let $(\cdot,\cdot)$ denote the L^2 inner product on Ω. Applying (9), we rewrite (12) as

$$\psi\frac{\partial\bar{C}}{\partial t} + \mathbf{u}\cdot\nabla\bar{C} \equiv \sqrt{\psi^2 + |\mathbf{u}|^2}\frac{\partial\bar{C}}{\partial\tau} = \nabla\cdot D\nabla\bar{C}, \tag{13}$$

where $\frac{\partial\bar{C}}{\partial\tau}$ is a directional derivative. Denote by $\bar{C}_h^{n+1} \in \mathcal{M}_h$ the approximation to $\bar{C}(t^{n+1})$. We determine $\bar{C}_h^{n+1}$ by

$$(\psi\frac{\bar{C}_h^{n+1} - \bar{\bar{C}}_h^n}{\Delta t}, v) + (D\nabla\bar{C}_h^{n+1}, \nabla v) = 0, \quad v \in \mathcal{M}_h, \tag{14}$$

where

$$\bar{\bar{C}}_h^n(\mathbf{x}) = C_h^*(\mathbf{x} - \frac{\mathbf{u}(\mathbf{x})}{\psi(\mathbf{x})}\Delta t, t^{n+1}). \tag{15}$$

Note that (14) is obtained from (13) by integrating by parts and approximating the τ derivative using backward differencing in time.

For more details on this time-splitting procedure, and for theoretical error estimates, see [30]. The MMOC has been analyzed extensively for transport problems, see [5] and the references therein, and comparisons with analytical solutions can be found in [4].

One major advantage of the MMOC over other transport procedures is that large time steps can be taken, without loss of accuracy or stability, because the time truncation error is $\mathcal{O}(\Delta t||c_{\tau\tau}||)$, where $||c_{\tau\tau}||$ is the L^2 norm of the second derivative of the solution in the τ-direction [10]. This term is generally much smaller than the time truncation error for standard second order in time procedures. By using time-splitting, we are able to approximate each piece of the calculation, namely, transport and biodegradation, with an appropriate numerical method, and an appropriate time step. The error analysis given in [30] for the time-splitting procedure discussed here shows that, for sufficiently smooth C, the L^2 norm of the error at any time level is $\mathcal{O}(h^2 + \Delta t + \Delta t_s^2)$.

4 Numerical results

The numerical method described above has been applied to the simulation of vertical cross-sectional flow. In these studies, we have concentrated on two aspects of the problem, namely, the effects of variable hydraulic conductivity, and the effects of spatially varying retardation. For simplicity, we consider the mathematical model of Borden and Bedient. Thus, we simulate the flow of dissolved oxygen and a single substrate, and the interaction of these components with a single microbial population.

Biodegradation in this model is assumed to satisfy the Monod kinetics, as given in the following system of ordinary differential equations:

$$R_S \frac{dS}{dt} = -B \cdot \bar{k} \cdot (\frac{S}{K_S + S})(\frac{O}{K_O + O}) \equiv RS(O, S, B), \tag{16}$$

$$\frac{dO}{dt} = -B \cdot \bar{k} \cdot F \cdot (\frac{S}{K_S + S})(\frac{O}{K_O + O}) \equiv RO(O, S, B), \tag{17}$$

and

$$\begin{aligned} \frac{dB}{dt} &= B \cdot \bar{k} \cdot Y \cdot (\frac{S}{K_S + S})(\frac{O}{K_O + O}) - b \cdot B + B_g \\ &\equiv RB(O, S, B), \end{aligned} \tag{18}$$

where $\bar{k}$ = maximum hydrocarbon utilization rate per unit mass microorganisms, Y = microbial yield coefficient, K_S = hydrocarbon half saturation constant, K_O = dissolved oxygen half saturation constant, b = microbial decay rate, and F = ratio of dissolved oxygen to hydrocarbon consumed. The last term on the right side of (18), B_g, is a small growth term which assures that microbes do not become extinct. Combining (16)-(18) with (5), (6), (8), and (9) completes the definition of the mathematical model.

As seen in (5), the assumption of linear, instantaneous adsorption gives rise to a retardation factor for substrate, which is given by

$$R_S = 1 + \rho_b K_d / \phi, \tag{19}$$

where ρ_b is the bulk mass density (g/cm^3), and K_d is the distribution coefficient (cm^3/g), see [12]. The latter term can be expressed as

$$K_d = K_{oc} f_{oc}, \tag{20}$$

where K_{oc} is the soil partition coefficient normalized to the organic carbon content of the soil, and f_{oc} is the fraction of organic carbon in the soil.

Field studies indicate that f_{oc} may vary between .01% and 1% within a field site [3], thus f_{oc} varies spatially, sometimes by orders of magnitude. Since no precise spatial representation for f_{oc} is known, we will make three different assumptions on the spatial distribution of organic carbon:

1) f_{oc} is uniform.

2) The values of f_{oc} satisfy certain statistical properties; f_{oc} values at various spatial locations were generated stochastically using a code developed by D. Moissis. In this code, the user specifies correlation length, mean, and standard deviation. A spatially correlated, log-normal distribution of values is then generated. Finally, a small amount of random noise is added. For more details on this algorithm, see [19].

3) Values of f_{oc} decrease with depth. This assumption is based on the fact that organic carbon content is proportional to the amount of humic substance in the soil, which is more prevalent near the surface.

The expected effect of varying f_{oc} spatially is to induce another level of variation into the effective velocity for the substrate.

Karickhoff, *et al* [13], showed that reasonable partition coefficients (K_{oc}) could be deduced from the octanol/water partition of the chemical, K_{ow}. They presented the following relationship between K_{oc} and K_{ow},

$$\log K_{oc} = \log K_{ow} - 0.21. \tag{21}$$

Values for K_{ow} can vary widely, depending on the substrate. For benzene, toluene, and xylene, K_{ow} varies from $135 - 1500\ cm^3/gm$ [18]. Moreover, for porosities in the .2-.4

range, ρ_b varies between $1.6 - 2.1\ g/cm^3$. Given these ranges of values for K_{ow}, ρ_b, and f_{oc}, the retardation factors for different compounds may vary between 1 and 129.

In the numerical simulations described below, the physical domain Ω is a rectangular cross-section of an aquifer, 260 feet long and 6 feet deep. Hydraulic head is specified on the left and right edges of Ω, and no-flow boundary conditions are assumed at the top and bottom; the head drop is 7.5 feet from left to right. Gravitational effects are not included, since density differences are negligible. In the transport equations, concentrations of dissolved oxygen and substrate are specified on the inflow (left) boundary, and at the right boundary we assume

$$\phi R_C C_t + \mathbf{u} \cdot \nabla C = 0, \quad C = O, S; \tag{22}$$

thus, substrate and dissolved oxygen are transported out of the domain through the right boundary. Substrate concentration is assumed to be zero at the inflow boundary, and dissolved oxygen concentration is given by

$$O(x=0,t) = \begin{cases} t/20 \text{ (days)} * 40 \text{ mg/l}, & 0 \le t \le 20, \\ 40 \text{ mg/l}, & \text{otherwise.} \end{cases} \tag{23}$$

Initially, no dissolved oxygen is present in the aquifer, and constant substrate and microbe concentrations of 6 and .002 mg/l, respectively, are assumed. Values for porosity, longitudinal and transverse dispersion, and the reaction coefficients are given in Table 1. These values remain constant throughout all of the simulations described below. Moreover, for all of the runs described below, except where noted, an 82 by 25 point-centered nonuniform mesh was employed, and the transport time step was .5 days, while the reaction time step was .00625 days.

In (19), we set $\rho_b = 2\ gm/cm^3$, and $K_{oc} = 259.5\ cm^3/g$. Corresponding to assumptions 1-3 on f_{oc} discussed above, we will consider five different functional forms for f_{oc}:

I. $f_{oc} \equiv 0$, thus $R_S \equiv 1$.

II. $f_{oc} \equiv .001$, giving $R_S \equiv 2.73$.

III. f_{oc} varying stochastically, with mean $1.2 * 10^{-4}$, standard deviation of $1.6 * 10^{-4}$, minimum of 0, and maximum of $1.34 * 10^{-3}$. This results in a retardation factor varying between 1 and 3.3, with mean 1.2. The correlation length in this case is roughly 10 feet.

IV. f_{oc} varying stochastically, with mean $9.6 * 10^{-4}$, standard deviation of $9.6 * 10^{-4}$, minimum of $3 * 10^{-4}$ and maximum of $7.9 * 10^{-3}$. This results in a retardation factor varying between 1.5 and 14.667, with mean value 2.7. Again, the correlation length is 10 feet.

V. f_{oc} varying linearly with depth, with $f_{oc} = .5\%$ at the top of the aquifer, and $f_{oc} = 0$ at the bottom. This leads to a retardation factor varying linearly between 1 and 9.7, and decreasing with depth. The arithmetic mean retardation factor is 5.35.

ϕ	.3
α_l	.6 ft
α_t	.001 ft
$\bar{k}$	1.7 /day
F	3
Y	.5
K_S	.13 mg/l
K_O	.1 mg/l
B_g	$2 * 10^{-5}$ mg/l/day
b	.01 /day

Table 1: Values of Physical Parameters.

We first consider a homogeneous medium, where $K \equiv 1.9 * 10^{-3}$ cm/sec, and demonstrate the effects on transport and biodegradation of varying retardation spatially. In Figures 1a through 1e, we have contoured substrate concentration at time 250 days, for f_{oc} as given in I through V, respectively. Examining these results, we see that larger values of R_S result in slower flow of substrate, and varying R_S spatially causes substantial perturbation of the concentration profile. In each case, a "reaction zone" develops, which moves with the flow, where O and S are nonzero, and biodegradation occurs. This zone is relatively narrow in these simulations, since line drive flow is modeled. For R_S values close to 1, the reaction zone is extremely thin at any fixed time, since dissolved oxygen and substrate move at roughly the same speed. In fact, the predominant mixing effect in this situation is physical dispersion. As R_S gets larger, the zone widens. Dissolved oxygen moves at a faster

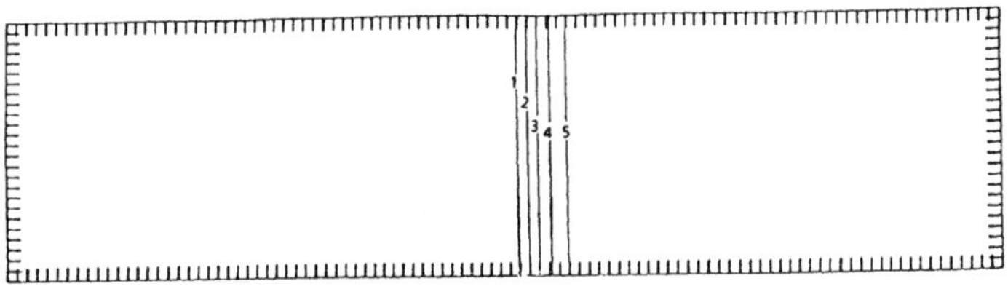

Figure 1a: Substrate concentration, mg/l

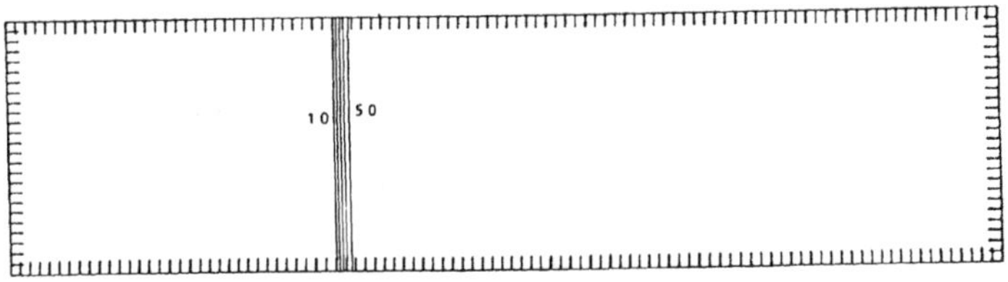

Figure 1b: Substrate concentration, mg/l

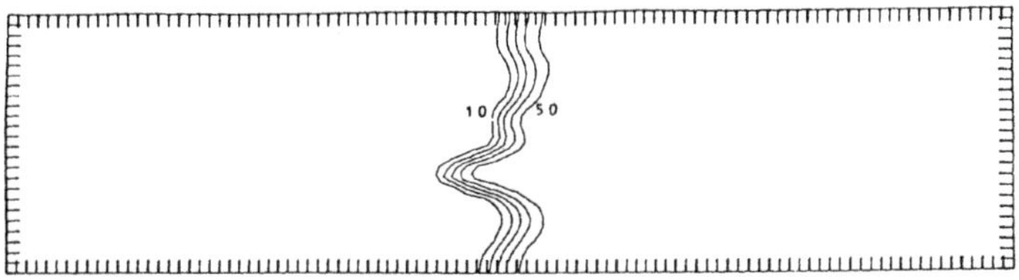

Figure 1c: Substrate concentration, mg/l

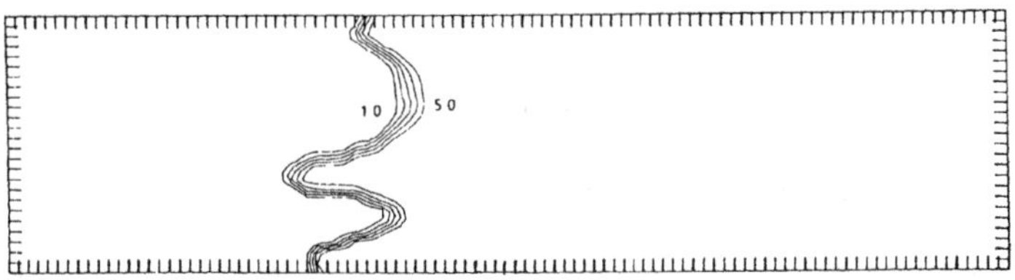

Figure 1d: Substrate concentration, mg/l

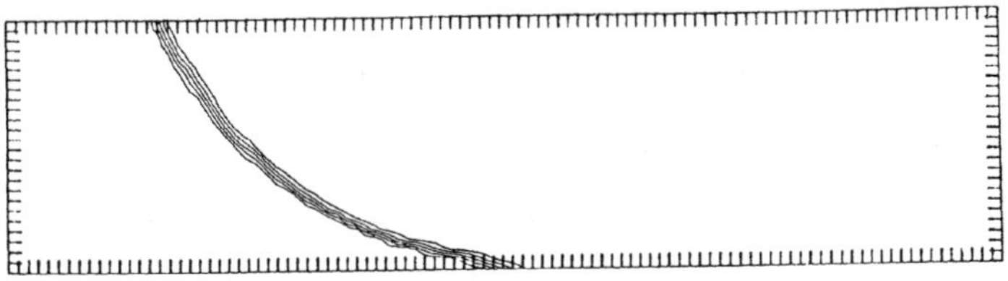

Figure 1e: Substrate concentration, mg/l

rate than substrate, and overtakes it, allowing for biodegradation to occur over a larger spatial region. However, the amount of biodegradation which can occur at any fixed point in space is actually reduced by increased adsorption. This dichotomy is reflected in (16) and will be examined in more detail below. We also note that, in the higher retardation cases II and IV, the substrate profile shows a steeper front than in the lower retardation cases I and III, compare Figures 1b and 1d with Figures 1a and 1c. The reasons for this are two-fold. First, effective dispersion (D/R_S) levels are lower in cases II and IV; moreover, more biodegradation is occurring at the leading edge of the front in these cases than in cases I and III, due to the increased overlapping of dissolved oxygen and substrate caused by retardation. Thus, the smooth edges of the front are removed, leaving a sharper profile. We will see this same effect in other situations below.

More information on these simulations is given in Figures 2 and 3. In Figure 2, we plot mass storage of substrate in solution at time t (days) for all of the cases listed above; that is, we are plotting

$$f(t) = \int_\Omega \phi S(x,t)dx.$$

Here, f is in milligrams of substrate. Note that, in all cases, f is a decreasing function of time, and f increases with increasing retardation factor. The decrease in f with time is due to biodegradation and flow of substrate out of Ω; recall that we assume no substrate flows into the aquifer, since this boundary is rich in dissolved oxygen. The increase in f with R_S shows that, in all cases, the slower outflow of substrate caused by increasing adsorption rates dominates any increase in biodegradation due to overlapping of dissolved oxygen and retarded substrate. This is obvious from comparing the position of the fronts in Figures 1a-1e. A closer look at the relationship between adsorption and biodegradation is given in

Figure 3. In this figure, we plot the amount of substrate mass biodegraded versus time, given by (see (16))

$$g(t) = -\int_0^t \int_\Omega \phi R_S^{-1} RS(O, S, B) dx dt.$$

In this equation we have two competing terms, R_S^{-1} and RS. For $R_S = 1$, the overlap between dissolved oxygen and substrate is very narrow, causing RS to be zero over a large portion of the space-time domain. For R_S increasing but still close to one, the reaction zone grows, enlarging the area over which RS is nonzero, and thus increasing g. Eventually though, R_S^{-1} dominates, that is, a threshold retardation factor is reached where adsorption severely limits the amount of substrate mass biodegraded. This is verified in Figure 3. For cases I-IV, g increases with retardation (at any fixed time); however, in case V, which has the highest mean retardation factor, we see a decrease in mass biodegraded. In fact, g in case V lies between the case III and case IV values. We found by numerical experimentation that, for this particular problem, g increases with retardation up to $R_S \approx 3$, where adsorption begins to dominate, and from this point on, mass biodegraded decreases with increasing retardation. In general, this threshold retardation factor will vary with the soil conditions.

We next consider a more realistic scenario, with flow through a vertically stratified cross-section of a porous medium. A hydraulic conductivity field was generated which is representative of typical aquifers in the U. S. Gulf coastal region. In this field, the conductivity varies from $1 * 10^{-5}$ to $5 * 10^{-3}$ cm/sec, with arithmetic mean value of $1.9 * 10^{-3}$ cm/sec. A contour map of the hydraulic conductivity $K(x, y)$ is given in Figure 4. In Figure 5, we show the substrate solution at 125 days with $R_S \equiv 1$ and no aerobic biodegradation, that is, we assume no dissolved oxygen is present in the system. In Figure 6,

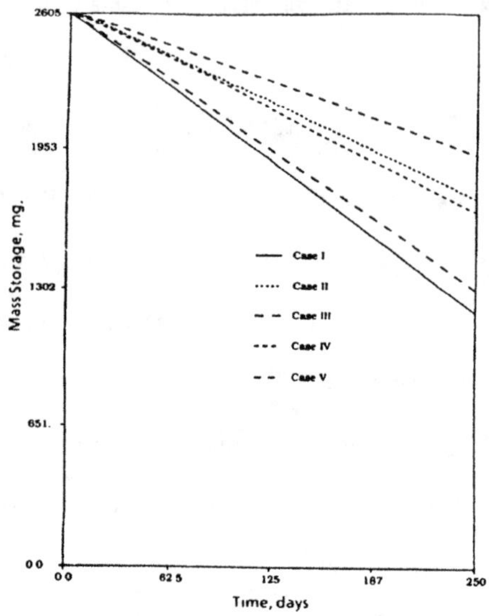

Figure 2: Mass storage (mg) vs. time (days) for cases I-V, $K \equiv 1.9 * 10^{-3}$ cm /sec.

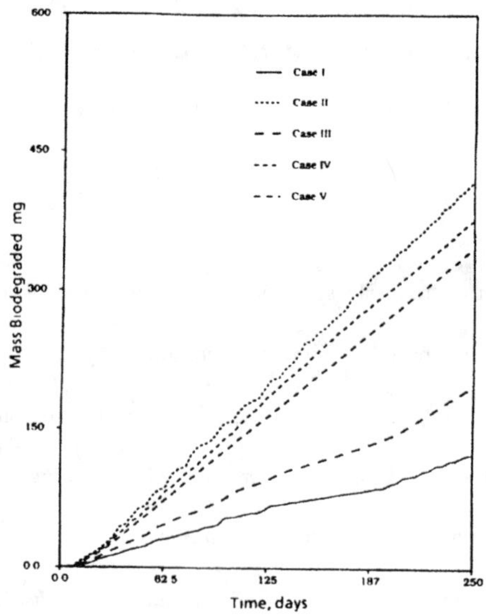

Figure 3: Mass biodegraded (mg) vs. time (days) for cases I-V, $K \equiv 1.9 * 10^{-3}$ cm /sec.

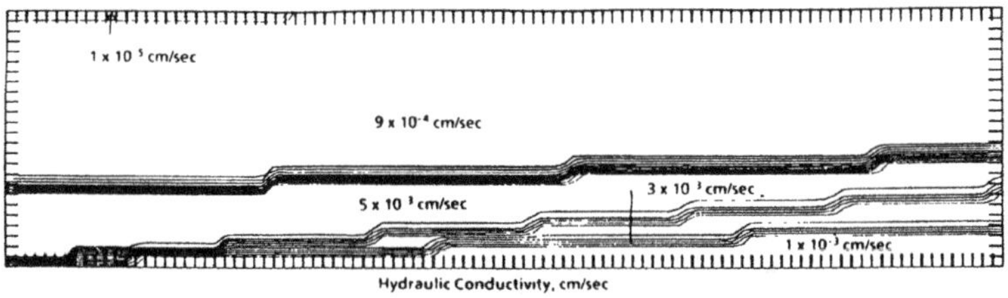

Figure 4: Contour map of vertically stratified conductivity.

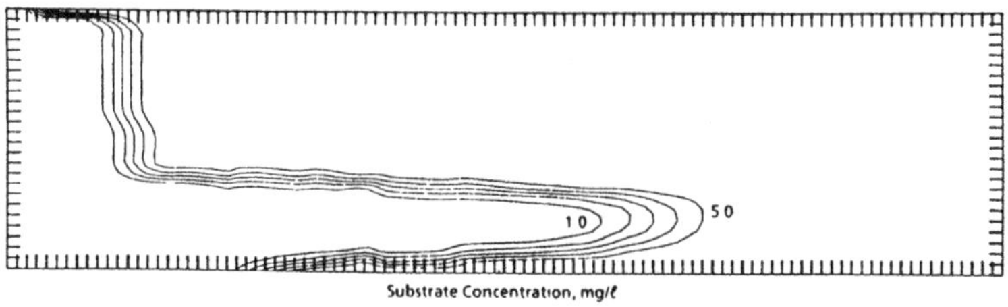

Figure 5: Substrate solution at 125 days, stratified medium, $R_S \equiv 1$, no biodegradation.

we show the same solution with biodegradation. Note that the protruding finger in these figures corresponds to flow in the higher hydraulic conductivity zone. Also, comparing Figures 5 and 6, we note that little biodegradation has occurred, since physical dispersion is low and dissolved oxygen and substrate are moving at the same speed. In Figure 7 and Figure 8, we do the same comparison for $R_S \equiv 2.73$, that is, Case II above. We see that a substantial amount of biodegradation has occurred. These results agree with the results of Figure 3 above. We also note that the biodegraded fronts are steeper than those without

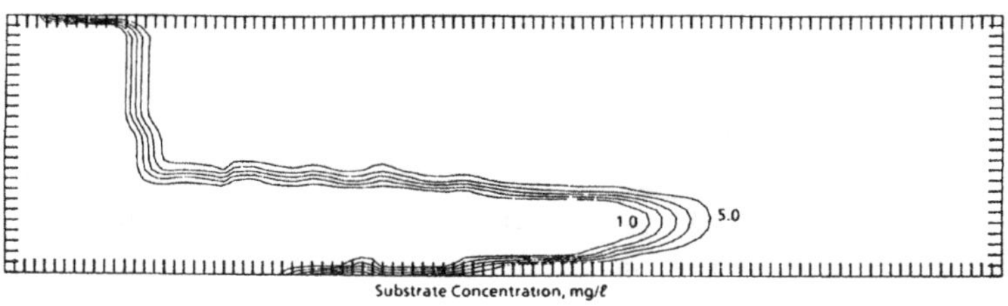

Figure 6: Substrate solution at 125 days, stratified medium, $R_S \equiv 1$, with biodegradation.

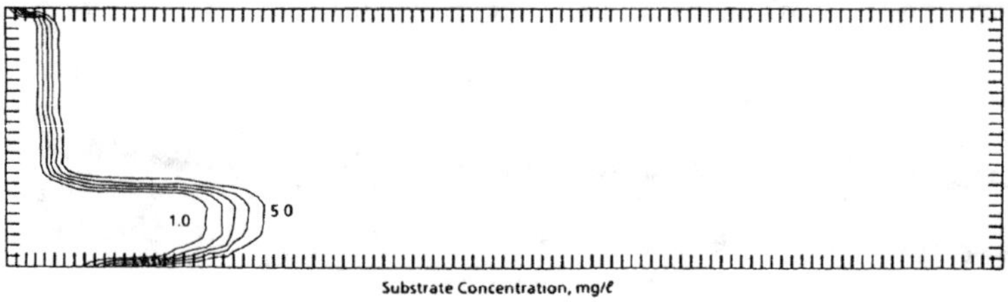

Figure 7: Substrate solution at 125 days, stratified medium, $R_S \equiv 2.73$, no biodegradation.

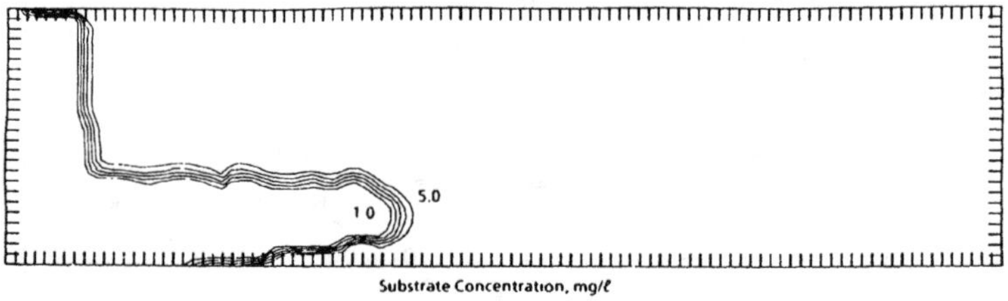

Figure 8: Substrate solution at 125 days, stratified medium, $R_S \equiv 2.73$, with biodegradation.

biodegradation.

Convergence of the substrate solution under mesh refinement is studied in Figures 9 and 10. Here we have taken the same physical data used to generate Figure 8 above and performed simulations with 82 by 48 and 91 by 70 nonuniform grids. As seen in these figures, the substrate front sharpens as the mesh is refined. However, the 82 by 48 and 91 by 70 solutions are virtually identical.

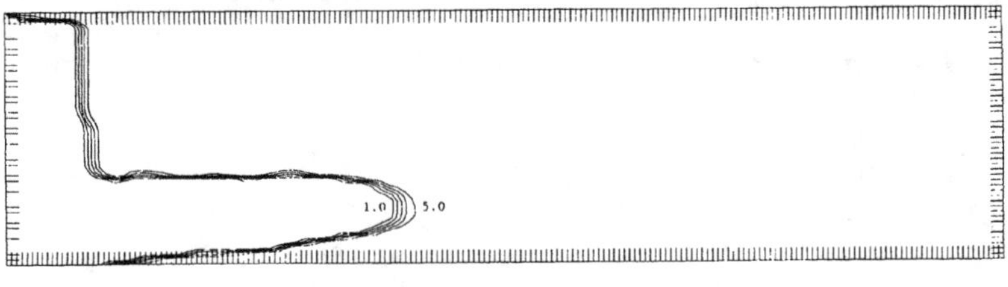

Figure 9: Substrate solution at 125 days, stratified medium, $R_S \equiv 2.73$, with biodegradation, 82 by 48 mesh.

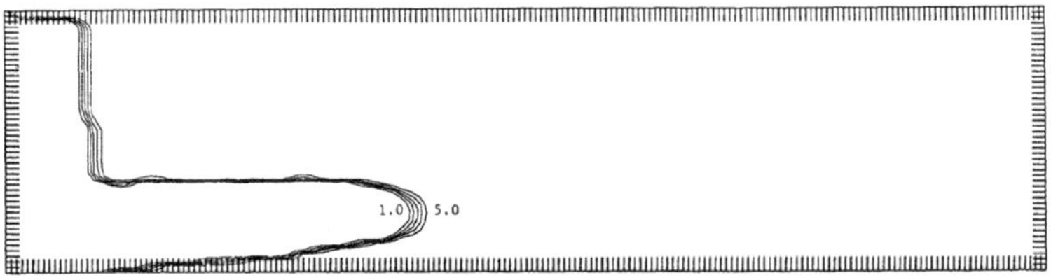

Substrate Concentration mg/ℓ

Figure 10: Substrate solution at 125 days, stratified medium, $R_S \equiv 2.73$, with biodegradation, 91 by 70 mesh.

The dissolved oxygen solution corresponding to the substrate solution in Figure 8 (stratified porous medium, $R_S \equiv 2.73$) is plotted in Figure 11. The figure shows that, as expected in this case, dissolved oxygen essentially fills the region void of substrate.

We have also considered the microbe solution for this case. Based on the mesh refinement study (Figures 8-10) above, the substrate solution for this problem is convergent on relatively coarse meshes. We have found by experimentation, however, that a truly accurate picture of the corresponding microbe solution requires very fine spatial mesh. In general, under the assumption of the Monod kinetics, the microbe solution is extremely sensitive to small variations in the hydrocarbon and oxygen solutions, at least for the values of reaction parameters chosen here. This sensitivity can be seen more clearly by the following argument. Holding substrate and dissolved oxygen concentrations fixed over a small time interval $[\tau, \tau+\epsilon]$, one can show, by solving (18) analytically, that,

$$B(x,y,\tau+\epsilon) = e^{\alpha(x,y)\epsilon}\left[\frac{B_g}{\alpha(x,y)} + B(x,y,\tau)\right] - \frac{B_g}{\alpha(x,y)}, \tag{24}$$

where

$$\alpha(x,y) = \bar{k} \cdot F \cdot \left(\frac{S(x,y)}{K_S + S(x,y)}\right) \cdot \left(\frac{O(x,y)}{K_O + O(x,y)}\right) - b.$$

Thus, for regions where $\alpha > 0$, which is usually the case when both substrate and dissolved oxygen are present, the solution B grows by an exponential factor. Thus, one can see that small variations in α can lead to large variations in the solution B.

In the stratified medium presented above, the predominant growth zones for microbes are the transition regions between higher and lower conductivity; these regions are also where α in (24) varies the most. The microbial growth is higher in these zones because the flow patterns dictate that substrate from the lower conductivity layers feeds into the higher conductivity layer, where dissolved oxygen is present. Eventually, the dissolved oxygen front in the lower conductivity zones catches up with substrate; however, in the transition regions α is positive for a substantial period of time, during which microbes grow exponentially. Since physical (especially transverse) dispersion in this case is assumed extremely low, the microbe solution exhibits very sharp "peaks" in these transition zones. This effect is exhibited in Figure 12. Here we plot the microbe solution at 125 days for the stratified porous medium with constant $R_S = 2.73$, at $x = 40$ feet, for three different meshes. The meshes used were 119 by 97, 159 by 114, and 164 by 155, with most of the grid blocks concentrated in and around the higher conductivity layer. The solutions exhibit two very sharp peaks, corresponding to the two transition zones between higher and lower conductivity (see Figure 4). The peak becomes sharper with each mesh refinement, but the solutions appear to be converging. Inside the higher conductivity layer, more moderate microbial growth occurs, while outside these regions, microbes are essentially dormant. This phenomenon of microbial growth in transition zones between higher and lower hydraulic conductivity layers has also been observed in the laboratory [29].

In Figures 13, 14, and 15, we have plotted the substrate solutions at 125 days for the stratified porous medium, with f_{oc} as given in III-V, respectively. Comparing these results

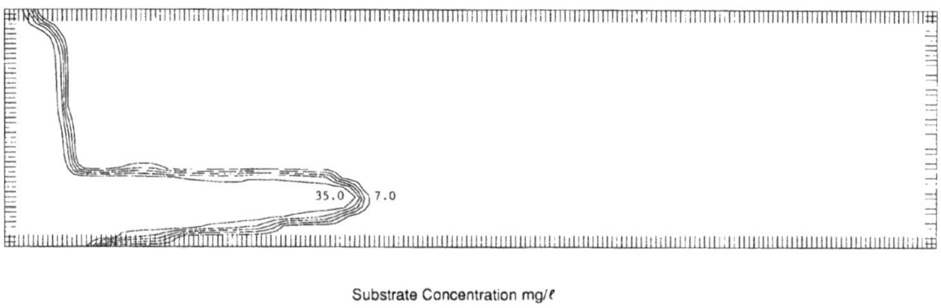

Figure 11: Dissolved oxygen solution at 125 days, corresponding to substrate solution in Figure 8, stratified medium, $R_S \equiv 2.73$.

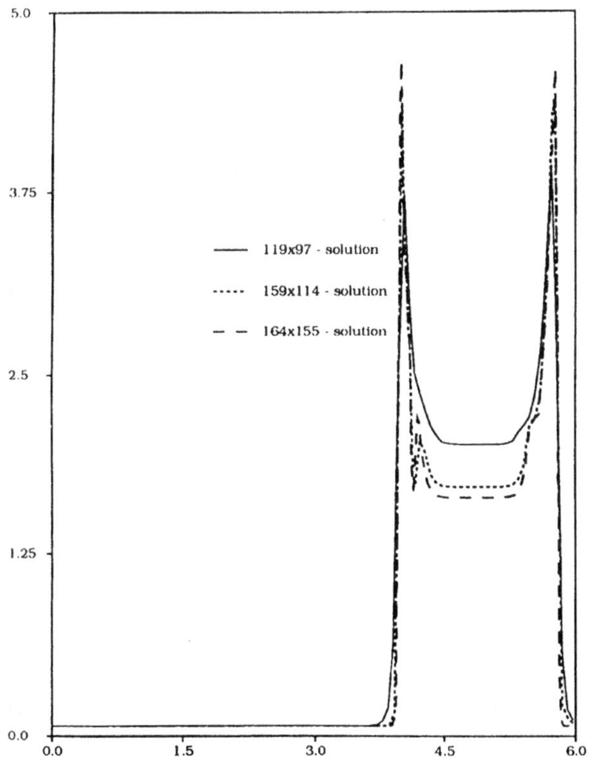

Figure 12: Microbe solution along $x = 40$ feet at 125 days for three different meshes, stratified porous medium, $R_S \equiv 2.73$.

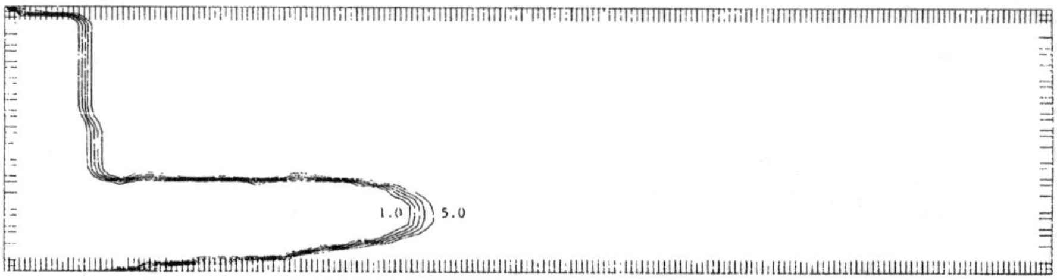

Figure 13: Substrate solution at 125 days, stratified medium, variable f_{oc}, mean $R_S = 1.2$.

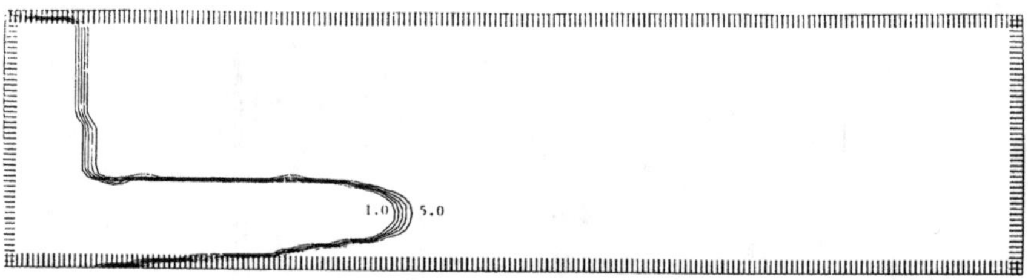

Figure 14: Substrate solution at 125 days, stratified medium, variable f_{oc}, mean $R_S = 2.7$.

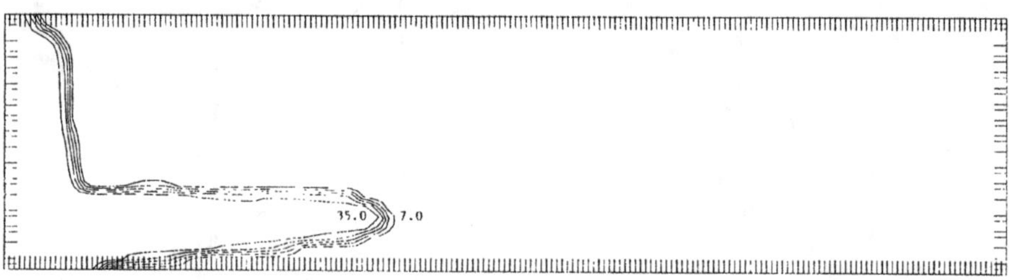

Figure 15: Substrate solution at 125 days, stratified medium, stratified f_{oc}.

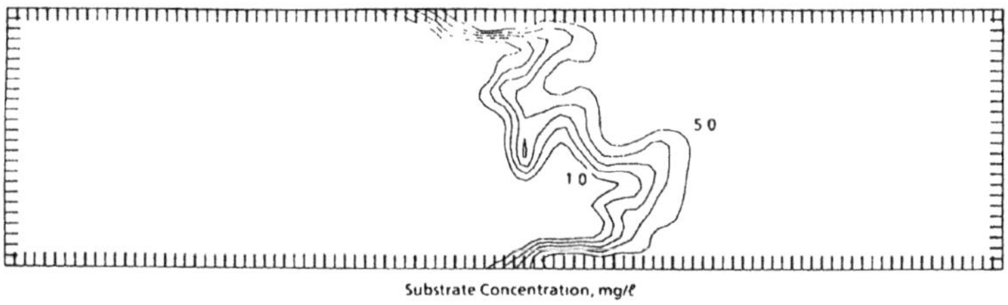

Figure 16: Substrate solution at 250 days, K varying statistically, constant f_{oc}, $R_S \equiv 1$.

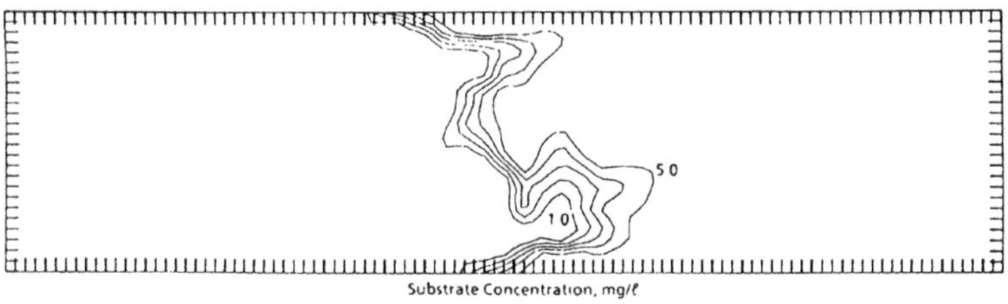

Figure 17: Substrate solution at 250 days, K varying statistically, variable f_{oc}, mean $R_S = 1.2$.

to those given in Figures 6 and 8, we see that, for these runs, the dominant effect on the flow is the high conductivity layer.

In our final set of simulations, we consider a conductivity field $K(x, y)$ generated statistically, again using the Moissis code. In these runs, the statistical mean of K is $1.9 * 10^{-3}$ cm/sec, the minimum value is $3.9 * 10^{-4}$, the maximum value is $8.1 * 10^{-2}$, and the standard deviation is $3.07 * 10^{-3}$ cm/sec. The correlation length is 10 feet in the longitudinal direction and .24 feet in the transverse direction. We will again examine the relative effects of conductivity and organic carbon variations. First, the substrate solution for $f_{oc} = 0$ at 250 days is plotted in Figure 16. We compare this solution to Figure 17, which corresponds to statistically varying f_{oc} as described in III above, Next, the substrate solution for $f_{oc} \equiv .001$ at 250 days is plotted in Figure 18. We compare this to the substrate solution in Figure 19, which corresponds to IV above. Note that, while differences in the solutions are definitely

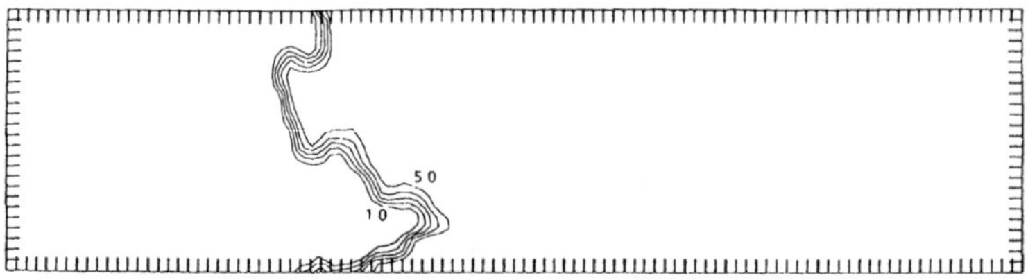

Figure 18: Substrate solution at 250 days, K varying statistically, constant f_{oc}, $R_S = 2.73$.

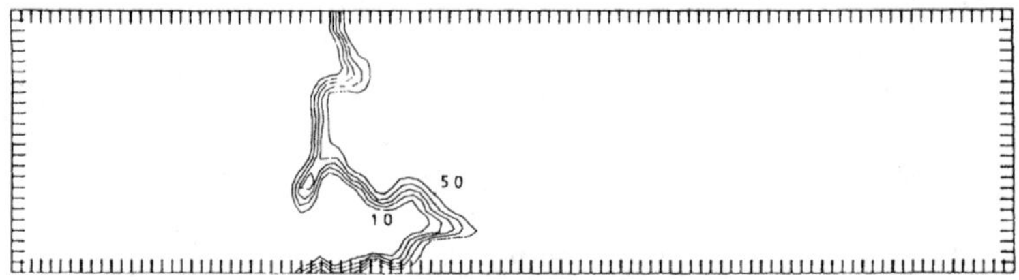

Figure 19: Substrate solution at 250 days, K varying statistically, variable f_{oc}, mean $R_S = 2.7$.

observed, qualitatively the solutions are very similar. Thus, conductivity variations still seem to be dominating factor. We now consider stratified f_{oc} (V). In Figures 20-22, we show the substrate solutions at 250, 375, and 500 days for this case. We see that these results are qualitatively similar to the result in Figure 1e above, where conductivity was constant. Thus, retardation variations seem to dominate in this situation.

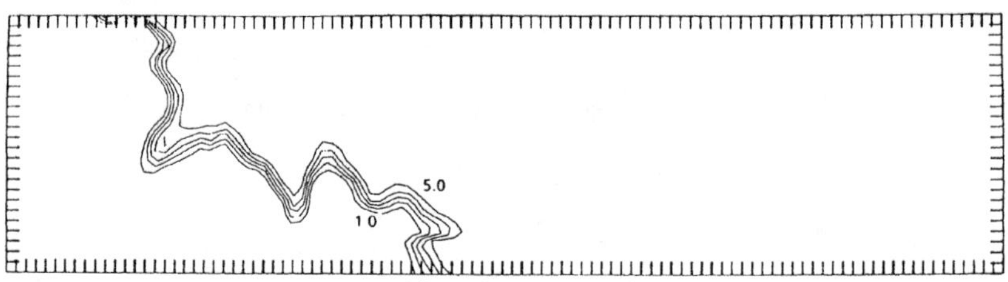

Figure 20: Substrate solution, K varying statistically, stratified f_{oc}, 250 days.

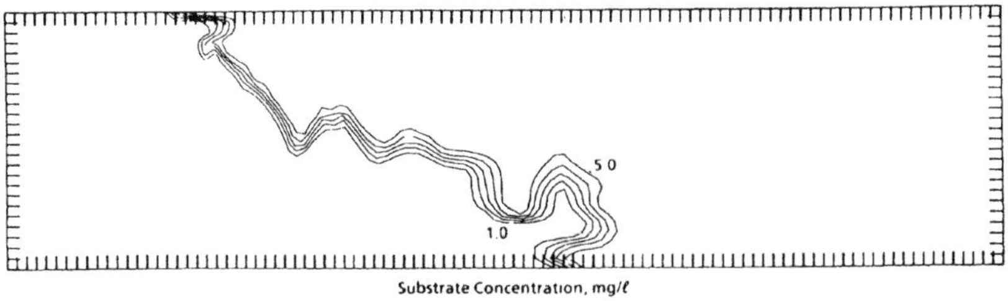

Figure 21: Substrate solution, K varying statistically, stratified f_{oc}, 375 days.

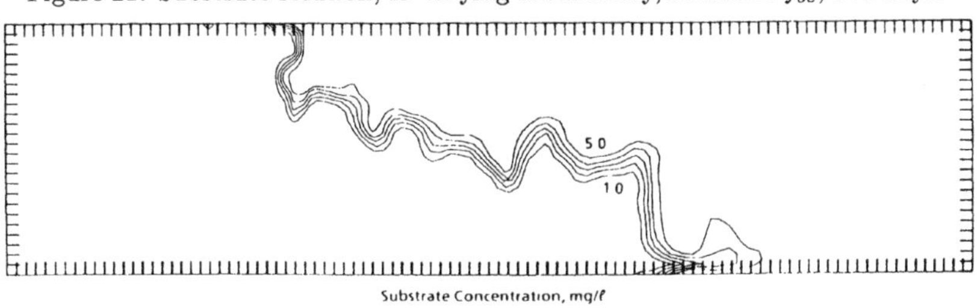

Figure 22: Substrate solution, K varying statistically, stratified f_{oc}, 500 days.

5 Conclusions

An efficient numerical method has been presented which can be used to study the fate of soluble organics under the influence of enhanced biorestoration. This method treats systems of transport-biodegradation equations by operator splitting in time. Transport is approximated by a finite element modified method of characterisics. The biodegradation terms are split from the transport terms and treated as a system of ordinary differential equations. 28 The method has been applied to a typical enhanced biorestoration scenario, where dissolved oxygen is introduced into a contaminated aquifer. The effects of spatially varied hydraulic conductivity and organic carbon content (sorption) on the fate of soluble organics has been studied. We now summarize the results of this study, which are particular to the cases considered and to the mathematical model, but may also give insight to other practical problems of interest.

We assumed an aquifer uniformly rich in organic carbon and substrate, and introduced dissolved oxygen to the system through an inflow boundary. Thus, a reaction zone developed, whose width depended on the levels of diffusion, dispersion, and substrate retardation. We assumed fixed diffusion and dispersion, and studied only the effects of retardation. When the retardation factor was high, dissolved oxygen moved at a faster rate than substrate, and overtook it, allowing for biodegradation to occur over a larger spatial region. In addition, sharper substrate concentration fronts were observed. The following two mechanisms caused these sharp fronts to develop. First, effective dispersion (D/R_S) was decreased with increased retardation. Second, due to the increased overlapping of dissolved oxygen and substrate, the smooth edges of the substrate front were removed by biodegradation, leaving a sharper profile.

The simulations on the interactions between adsorption and biodegradation showed that, within a certain range, higher adsorption rates resulted in more substrate mass biodegraded. However, when the retardation factor increased above a threshold level, the substrate mass biodegraded decreased with increasing adsorption rates. This threshold level should vary with the soil conditions. When the sorption rate increased past this level, a significant amount of substrate mass was adsorbed and became unavailable to microorganisms.

Based on several numerical experiments, spatially varying hydraulic conductivity and organic carbon content had significant influence on solute transport. Not surprisingly, however, the dominant effect on the distribution of soluble organics in most cases was the spatial pattern of the hydraulic conductivity field.

Finally, in a vertically stratified porous medium with a higher conductivity layer sandwiched between regions of lower conductivity, the growth of microorganisms was predominant in the transition areas between higher and lower hydraulic conductivity. This occurred because dissolved oxygen quickly filled the higher conductivity layer, and substrate was continually fed into this layer from the regions of lower conductivity.

Acknowledgments: The second author acknowledges the support of the National Science Foundation, Grant No. DMS-8807257, the National Center for Supercomputing Applications (NCSA) at the University of Illinois in Urbana-Champaign, and the National Science Foundation Center for Research in Parallel Computation. Some of the computer simulations described above were run on the CRAY X-MP/48 at NCSA. The third author acknowledges the support of the Department of Energy, Grant No. DE-FG05-88ER-25060. The authors would also like to acknowledge the contributions of David Moissis and Mandri Obeyesekere to the preparation of this work.

References

[1] Borden, R. C. and P. B. Bedient, *Transport of dissolved hydrocarbons influenced by oxygen-limited biodegradation 1. Theoretical development*, Water Resour. Res. 22, pp. 1973-1982, 1986.

[2] Celia, M. A., J. S. Kindred, and I. Herrera, *Contaminant transport and biodegradation, 1. A numerical model for reactive transport in porous media*, Water Resour. Res. 25, pp. 1141-1148, 1989.

[3] Chiang, C. Y., J. P. Salanitro, E. Y. Chai, J. D. Colthart, and C. L. Klein, *Aerobic biodegradation of benzene, toluene, and xylene in a sandy aquifer-data analysis and*

computer modelling, Ground Water 27, pp. 823-834, 1989.

[4] Chiang, C. Y., M. F. Wheeler, and P. B. Bedient, *A modified method of characteristics technique and a mixed finite element method for simulation of groundwater solute transport*, Water Resour. Res. 25, pp. 1541-1549, 1989.

[5] Dawson, C. N., T. F. Russell, and M. F. Wheeler, *Some improved error estimates for the modified method of characteristics*, SIAM J. Numer. Anal. 26, pp. 1487-1512 (1989).

[6] Dawson, C. N., M. F. Wheeler and P. B. Bedient, *Numerical modeling of subsurface contaminant transport with biodegradation kinetics*, Proceedings of National Water Wells Meeting, Houston, Texas, 1987, pp. 329-344.

[7] Dawson, C. N., M. F. Wheeler, T. M. Nguyen, and S. W. Poole, *Simulation of hydrocarbon biodegradation in groundwater*, CRAY Channels 8, 3, pp. 14-19, 1986.

[8] Dawson, C. N., M. F. Wheeler, T. M. Nguyen, and S. W. Poole, *Simulation of subsurface contaminant transport with biodegradation kinetics*, Proceedings of Third International Symposium on Science and Engineering on CRAY Supercomputers, Mendota Heights, Minnesota, pp. 75-86, 1987.

[9] Douglas, J. Jr., R. E. Ewing, and M. F. Wheeler, *Approximation of the pressure by a mixed finite element method in the simulation of miscible displacement*, RAIRO Anal. 17, pp. 17-33, 1983.

[10] Douglas, J. Jr. and T. F. Russell, *Numerical methods for convection-dominated diffusion problems based on combining the method of characteristics with finite element or finite difference procedures*, SIAM J. Numer. Anal. 19, pp. 871-885, 1982.

[11] Douglas, J., T. F. Dupont, H. H. Rachford, and M. F. Wheeler, *Local* H^{-1} *Galerkin and adjoint local* H^{-1} *Galerkin procedures for elliptic equations*, RAIRO Anal. Numer. 11, pp. 3-12, 1977.

[12] Freeze, R. A. and J. A. Cherry, Groundwater, Prentice-Hall, Englewood Cliffs, New Jersey, 1979.

[13] Karickhoff, S. W., D. S. Brown, and T. A. Scott, *Sorption of hydrophobic pollutants on natural sediments*, Water Research 13, pp. 241-248, 1979.

[14] Kindred, J. S. and M. A. Celia, *Contaminant transport and biodegradation 2. Conceptual model and test simulations*, Water Resour. Res. 25, pp. 1149-1159, 1989.

[15] Konikow, L. F. and J. D. Bredeheoft, *Computer model of two-dimensional solute transport and dispersion in groundwater. Automated data processing and computations.* Techniques of Water Resources Investigations of the United States Geological Survey, Washington, D. C., 1978.

[16] MacQuarrie, K. T. B., E. A. Sudicky, and E. O. Frind, *Simulation of biodegradable organic contaminants in groundwater 1, numerical formulation in principal directions*, Water Resour. Res. 26, pp. 207-222.

[17] MacQuarrie, K. T. B. and E. A. Sudicky, *Simulation of biodegradable organic contaminants in groundwater 2, plume behavior in uniform and random flow fields*, Water Resour. Res. 26, pp. 223-240.

[18] Miller, M. M., S. P. Wasik, G. L. Huang, W. Y. Shiu, and D. Mackay, *Relationship between octanol-water partition coefficient and aqueous solubility*, Environmental Science and Technology 19, pp. 522-529, 1985.

[19] Moissis, David, *Simulation of viscous fingering during miscible displacement in nonuniform porous media*, Ph. D. Thesis, Rice University, 1988.

[20] Molz, F. J., M. A. Widdowson, and L. D. Benefield, *Simulation of microbial growth dynamics coupled to nutrient and oxygen transport in porous media*, Water Resour. Res. 22, pp. 1207-1216, 1986.

[21] Nakata, M., A. Weiser, and M. F. Wheeler, *Some superconvergence results for mixed finite element methods for elliptic problems on rectangular domains*, MAFELAP Proceedings V, J. A. Whiteman, ed., Academic Press, New York, pp. 367-390, 1985.

[22] Raymond, R. L., Jamison, V. W., and Hudson, J. D., *Final report on beneficial stimulation of bacterial activity in ground water containing petroleum products*, Committee on Environmental Affairs, Americal Petroleum Institute, Washington, D. C., 1975.

[23] Rifai, H. S., *Numerical techniques for modeling* in situ *biorestoration and biodegradation of organic contaminants in ground water*, Ph. D. Thesis, Rice University, 1989.

[24] Rittman, B. E. and P. L. McCarty, *Model of steady state biofilm kinetics*, Biotech. Bioeng. 22, pp. 2343-2357, 1980.

[25] Russell, T. F., *Time-stepping along characteristics with incomplete iteration for a Galerkin approximation of miscible displacement in porous media*, SIAM J. Numer. Anal. 22, pp. 970-1013, 1985.

[26] Russell, T. F., M. F. Wheeler, and C. Y. Chiang, *Large-scale simulation of miscible displacement*, Proceedings of SEG/SIAM/SPE Conference on Mathematical and Computational Methods in Seismic Exploration and Reservoir Modeling, W. E. Fitzgibbon, ed., Society for Industrial and Applied Mathematics, Philadelphia, pp. 85-107, 1986.

[27] Thomas, J. M., M. D. Lee, P. B. Bedient, R. C. Borden, L. W. Canter, and C. H. Ward, *Leaking underground storate tanks: remediation with emphasis on* in situ *biorestoration*, Environmental Protection Agency, 600/2-87,008, January, 1987.

[28] United States Department of Energy, *Site-directed subsurface environmental initiative, five year summary and plan for fundamental research in subsoils and in groundwater, FY1989-FY1993*, DOE/ER 034411, Office of Energy Research, April 1988.

[29] United States Department of Energy, Review of intermediate-scale experiments for subsurface microbiology and chemistry, DOE/ER-0383, Proceedings of an OHER/DOE Sponsored Workshop at University of Delaware's Virden Center, April 12-14, 1987.

[30] M. F. Wheeler and C. N. Dawson, *An operator-splitting method for advection-diffusion-reaction problems*, MAFELAP Proceedings VI, J. A. Whiteman, ed., Academic Press, pp. 463-482, 1988.

[31] M. F. Wheeler, C. N. Dawson, P. B. Bedient, C. Y. Chiang, R. C. Borden, and H. S. Rifai, *Numerical simulation of microbial biodegradation of hydrocarbons in groundwa-*

ter, Proceedings AGWSE/IGWMCH Conference on Solving Ground Water Problems with Models, National Water Wells Association, pp. 92-108, 1987.

[32] Widdowson, M. A., F. J. Molz, and L. D. Benefield, *Development and application of a model for simulating microbial growth dynamics coupled to nutrient and oxygen transport in porous media*, Proceedings AGWSE/IGWMCH Conference on Solving Ground Water Problems with Models, National Water Wells Association, pp. 28-51, 1987.

Transport in Porous Media **6**: 703–737, 1991
© 1991 *Kluwer Academic Publishers. Printed in the Netherlands.*

REACTION KINETICS AND TRANSPORT IN SOIL: COMPATIBILITY AND DIFFERENCES BETWEEN SOME SIMPLE MODELS

Sjoerd E.A.T.M. van der Zee

Department of Soil Science and Plant Nutrition,

Agricultural University, Wageningen, The Netherlands

Short title: Reaction Kinetics and transport in soil

ABSTRACT

Non-equilibrium between the composition of the soil solution and of the sorption sites on the solid matrix of soil may be due to mass transfer resistances and to slow chemical kinetics of elementary or complex reactions. Different mechanistic models have been used to describe the rate of change of the soil solution composition due to sorption, using a good description of experimental results to draw conclusions with respect to the mechanisms involved. Assuming a mono-solute description to be appropriate, some similarities and differences between overall first order kinetics and two mechanistic diffusion-sorption kinetics models are considered. Empirical first order reaction rates can often be fitted well within experimental error on the solutions for the mechanistic diffusion-sorption models. For one of the mechanistic models (unreacted shrinking core) the behaviour is shown when the reactive solid particle size is exponentially distributed. This model can be recast into a semi-empirical model that allows scaling of reaction time, as its concentration dependence differs from the two other models. This is illustrated with an example taken from metallurgical engineering. Another difference is the effect of nonequilibrium on the solute front in an ideal plug flow reactor. For sorption according to the shrinking core model displacement may lead to a non-Fickian front, as front spreading does not grow with the square root of time. For the two other sorption models Fickian type of displacement results for large enough displacement distances.

Keywords: Kinetics, Diffusion, Shrinking core, Exposure, Scaling.

INTRODUCTION

Important aspects of chemical reactions, where reactants are transformed into products, are the stoichiometry, thermodynamic equilibrium and the kinetics of the reaction. In complex reactions several states can be distinguished between the initial reactant and final product composition. When no such intermediary states can be recognized a reaction is elementary, in a kinetic sense. In the latter case the reaction order, with respect to a component, is the power of the component concentration in the corresponding reaction rate expression. The reaction order equals the sum of all such powers of all reactants and equals, for mass action kinetics, the sum of stoichiometry coefficients of the reactants. For the elementary reaction

$$aA + bB \rightarrow cC + dD, \tag{1}$$

the reaction order equals a + b according to this convention.

For reactions occurring in soil, it is highly improbable that we deal with elementary reactions. Rather an overall equilibration rate is measured, comprising of parallel, series and/or complex reactions of dissolved reactants with different (non-pure) solid phases (e.g. clay colloids, sesquioxides, organic matter). These solid phases, even when present in pure form, are characterized by heterogeneity of the solid surface [Jaroniec, 1983, Van Riemsdijk et al., 1987, Hiemstra et al., 1989]. Hence, spectra of equilibrium and rate parameters rather than single valued realizations, are expected for reactions in soil. By appropriate macroscopic averaging this microscopic heterogeneity may be dealt with, sometimes even in closed form [Van Riemsdijk et al., 1987].

When we consider reactions with isolated uniform reactive soil compounds, we do not expect that these are elementary reactions, because in addition to chemical kinetics also physical factors lead to non-equilibrium. Envisioning a soil as an ensemble of aggregates, containing particles that in turn are constituted of different (reactive and non-reactive) grains, solutes may have to diffuse from the bulk of the solution outside the aggregates towards the reactive grains, etc. (Figure 1). This diffusional resistance implies non-equilibrium, even in the absence of slow chemical reaction rates. The accessibility of solid phase particles and grains depends on the mineral phase and the solute [Pfeiffer et al., 1983, Avnir et al.,

1984] as the surface is operationally defined. Therefore, when particles are not porous for the solute considered and aggregates are not present, diffusional resistances to sorption are limited to the transfer of solute through stagnant fluid films surrounding the solid phase. In general, when nonequilibrium in soil is described with a single rate parameter, physical and chemical resistances to equilibrium are lumped.

In view of the above, reaction rates in soil systems can be described in terms of mechanistic n'th order reactions in only a few cases, and for simple soil analogues. Empirical relationships may be employed, which have very limited extrapolative capabilities. Likewise, for a proper description in terms of diffusion, all resistances should be identified, as different rate and diffusion coefficients depend differently on system parameters. Because sorption studies in soil science are commonly based on the loss (rates) of solute from the bulk of the solution, this is taken as the point of departure. Even when simple mechanistic models are assumed to be appropriate, discrimination between these models using single curves for the solute loss in the solution as a function of time is shown to be difficult. Extrapolation of descriptions, e.g. in simulating transport, may yield completely different results, due to differences in concentration dependence of the models.

REACTION RATE AND DIFFUSION MODELS

Due to its simplicity, and because it often describes data well, the first order kinetics model has found wide use in soil science. This model may be appropriate for a first order chemical reaction rate as well as for mass transfer mechanisms. Likewise, overall sorption kinetics have been described with diffusion models (for diffusion into particles). These two models are discussed along with the unreacted shrinking core model, which has found use in chemical engineering, but rarely in soil science.

1. Apparent first order kinetics

For an elementary reaction of solute A with a solid surface (unit surface area), which is first order with respect to the reactant concentration at the liquid/solid interface, the reaction rate is

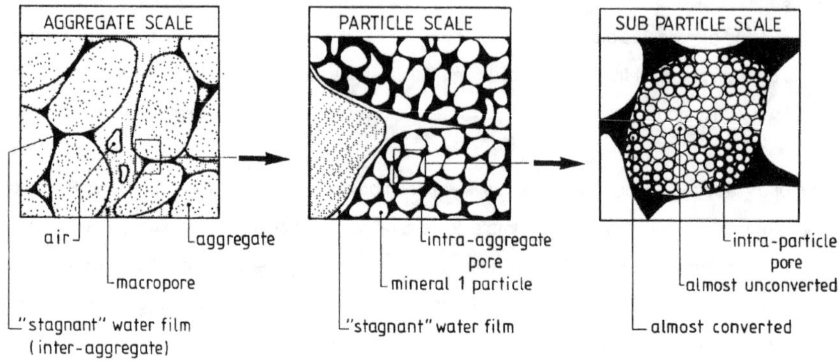

Figure 1: Illustration of multiple scales of diffusion in soil. Solute that arrived through the pores conducting water may diffuse into aggregates through a stagnant water film. Through intra-aggregates pores the solute may arrive at the solid particle surface. If such a particle is microporous and consists (partly) of reactive minerals, these may react and be converted.

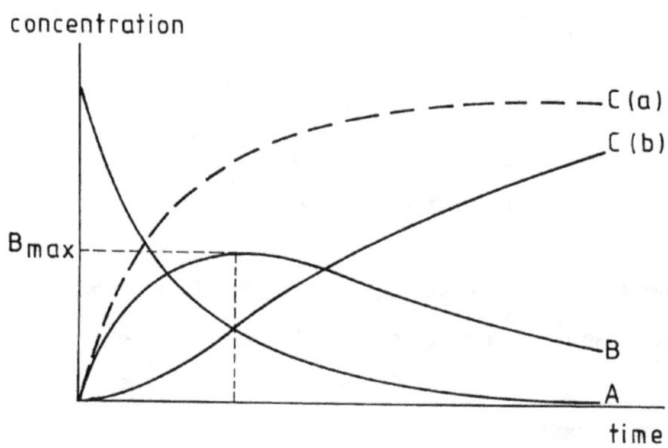

Figure 2: Concentration development in time for first order reaction rates. Elementary reaction (a) and reaction in series according to (5), with $k_{r2} = k_{r1}$ (b).

$$r_{A,i} = k_r \tilde{c}_{A,i} \tag{2}$$

for a unit volume of solution (For symbols: see Notation). Assume, that the concentration at the interface is replenished by diffusional mass transfer from the bulk of the solution [Bird et al., 1960]:

$$j_A = -k_f(\tilde{c}_{A,i} - c_A). \tag{3}$$

where A denotes the compound and i denotes the interface. The overall rate of disappearance in the bulk solution, which is usually the measured property, is found by combining (1) and (2). We assume a pseudo steady state, hence

$$r_A = \frac{k_r k_f}{k_r + k_f} c_A = \bar{k} c_A \tag{4}$$

This overall rate equation shows that we can not distinguish between a first order mass transfer or first order chemical reaction rate by measuring only c_A. Additional assumptions are needed with respect to the value or dependency on external factors (e.g. temperature) of k_r and k_f. The rate controlling mechanism defined on the basis of k-values, is the mechanism with the smallest rate constant, if k-values differ significantly and the mechanisms occur in series as in the above example.

As is clear from (4) first order kinetics may describe chemical and physical mechanisms, hence a good fit does not imply a mechanism. As was shown by Weber and Miller [1986, 1988] experimental results can often be described equally well with the first order model and with other models, such as those described in the next sections, for a limited range of time, temperature and concentration. Other (empirical) models may also provide a good description [Van Riemsdijk, 1979]. Changing the reaction period or concentration range of interest often required significant changes in model parameters or even sorption kinetics models. Again, in view of e.g. (4), a good fit of a model has little bearing on mechanisms active in overall sorption. A second example of where problems may be encountered is when the intermediary steps of the chemical reaction are not known. This is commonly the case when the nonequilibrium process is evaluated from the rate of solute loss in the solution. Considering the first order reactions in series given by

$$A \rightarrow B \rightarrow C \tag{5}$$

with initially $c_A = c_0$, $c_B = c_C = 0$, we find

$$c_A(t) = c_0 \exp(-k_{r1} t) \tag{6}$$

$$c_C(t) = c_0 \{1 - \frac{k_{r2}}{(k_{r2} - k_{r1})} \exp(-k_{r1} t) + \frac{k_{r1}}{(k_{r2} - k_{r1})} \exp(-k_{r2} t) \tag{7}$$

where the subscript numbers of k-values refer to the first and second step of (5). While the rate of solute loss (of A) thus follows first order kinetics with rate constant k_{r1}, the rate of formation of C (e.g. sorbed solute) does not follow from the disappearance of A directly, unless $k_{r2} \gg k_{r1}$, $t \gg 1/k_{r2}$. Complications, such as shown in this simple example can lead to significant problems in interpretation of data, when only A is measured and the intermediary reaction step is not recognized. For solutes subject to micro-biological transformations, the transformation product (C) may be measured. When the initial period before C is produced becomes significant (Figure 2) this may be explained as a lag phase of the micro-organisms involved. When, on the other hand, only the loss of A in solution is measured, it may be uncertain whether A is slowly sorbed (giving the sorbed complex B) or is actually transformed subsequent to sorption to give C. To understand the actual process the appropriate mechanisms should be identified. For this simple case this involves at least the measurement of A and C in solution as a function of time.

In the next Sections we consider first order kinetics for elementary reactions in more detail, assuming this mechanism to occur together with mass transfer resistances. Parallel reactions are discussed in chemical kinetics handbooks such as [Levenspiel, 1972, Pannetier and Sonchay, 1967].

2. Homogeneous Conversion Diffusion Models (HCD)

Consider a solute that diffuses into an aggregate or mineral particle. Resistances involved in such diffusion are for example diffusion through a stagnant fluid film, and through intra-aggregate or intra-particle pores. In the absence of a reaction and assuming a diffusion coefficient in the spherical particles that is independent of position, the governing equations are for the particle

$$\epsilon \frac{\partial \bar{c}}{\partial t} = D_s \left[\frac{\partial^2 \bar{c}}{\partial r^2} + \frac{2}{r} \frac{\partial \bar{c}}{\partial r} \right], \qquad 0 \le r \le R_0 \tag{8a}$$

$$\bar{c} = 0, \qquad 0 \le r \le R_0, \qquad t = 0 \tag{8b}$$

$$D_s \frac{\partial \bar{c}}{\partial r} = k_f (c - \bar{c}), \qquad r = R_0, \qquad t > 0 \tag{8c}$$

$$\frac{\partial \bar{c}}{\partial r} = 0, \qquad r = 0, \qquad t > 0, \tag{8d}$$

where c is the concentration in the bulk of the solution. The mass balance equation is given by (8a) with $\bar{c}$ the concentration in the sphere, the initial condition is given by (8b), fluid film diffusion is described with first order mass transfer (8c) and a no-flux boundary (8d) is found in the centre of the sphere. This set of equations describes the process shown in Figure 3a. Solutions were given for the above and other initial and boundary conditions by Crank [1975] and Carslaw and Jaeger [1959]. In view of the later analyses it is of interest to present the solution in general terms when an elementary first order reaction occurs in the region $0 \le r \le R_0$ (in case of a parallel equilibrium linear adsorption reaction onto the pore walls, the effects may be discounted for in the diffusion coefficient D_s). The amount that has diffused at time, t, may be given generically as follows. In general, by **G** we denote a solution of (8) in the absence of a (kinetic) reaction, which may be written in the form of infinite series such as

$$\mathbf{G} = c\left\{1 - \sum_i f_i(r)\exp(-w_i t)\right\}, \tag{9}$$

where the terms f_i and w_i are independent of time and where w_i is independent also of r. Examples of the terms f_i, w_i and g_i (see 10) for different geometries were presented by Crank [1975, p. 332-336] and omitted here for brevity and because they are not needed for our analysis in Section 4. The diffusion rate for that case is of the form

$$\tilde{M}_t = \frac{d\tilde{M}}{dt} = c\sum_i g_i \exp(-w_i t), \tag{10}$$

where g_i differs for each term in the series and is independent of r and t. In the presence of a first order kinetic reaction within the particle at the left hand side of (8a) the term $\epsilon k_r \bar{c}$ has to be added, where k_r is the first order reaction rate parameter. For that case the diffusion rate equals

$$M_t = k_r \int_0^t \tilde{M}_t \exp(-k_r t)dt + \tilde{M}_t \exp(-k_r t), \tag{11}$$

and with $\tilde{M}_t$ given in (10). Integrated this yields for the total mass that has diffused into the particle at time, t,

$$M(t) = \int_0^t M_t dt \tag{12}$$

$$= c\sum_i g_i \left\{\frac{k_r t(k_r + w_i) - w_i \cdot \exp[-t(k_r + w_i)] + w_i}{(k_r + w_i)^2}\right\}. \tag{13}$$

In the presence of a first order reaction within the particle, steady state may be approached for large $k_r t$-values, and the loss rate becomes constant. When $k_r t$-value are small the absorption curve, M(t), approaches the absorption curve in case of no kinetic reaction for decreasing values of $k_r t$.

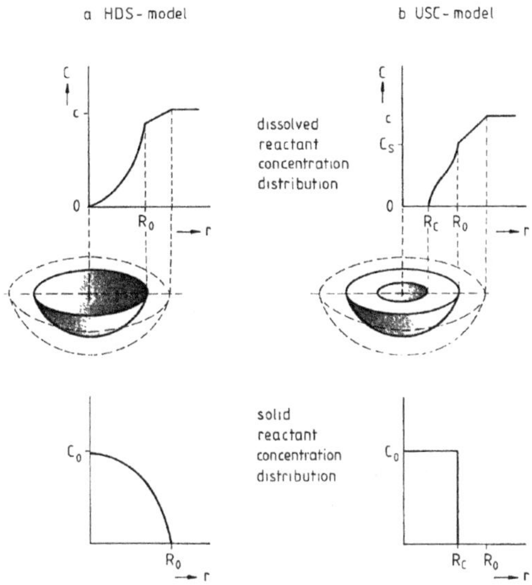

Figure 3: Schematic illustration of the homogeneous conversion diffusion (HCD:3a) and of the Unreacted Shrinking Core (USC:3b) models. Top gives reactant concentration and bottom gives the solid phase concentration.

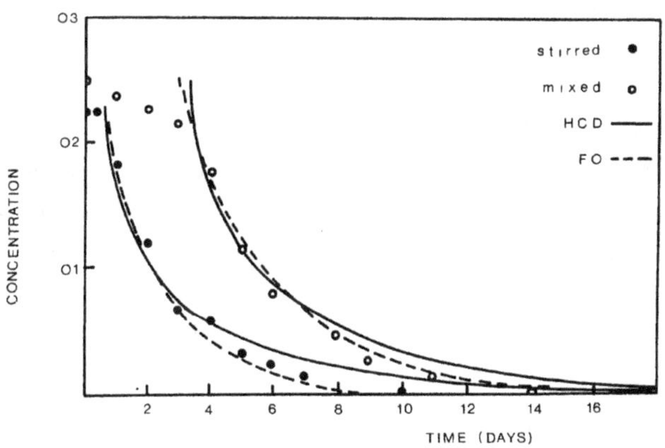

Figure 4: Fit of the first order (FO) kinetics and of the HCD models for α-HCH desorption and degradation in soil (after: Rijnaarts and Jumelet [1987]).

The diffusion model given by (8) considers more mechanisms explicitly than the apparent first order rate kinetics model. Because it has more adaptable parameters, it has a higher versatility to fit data. Increasing the number of parameters in the first order model, e.g. by assuming an equilibrium reaction in series or parallel with the kinetic process, it often becomes difficult to discriminate between the different models [Miller and Weber, 1986]. An example, taken from the conversion of $\alpha - HCH$ [Rijnaarts and Jumelet, 1987] shows equally good fits of the first order kinetics and of the HCD-model for desorption and degradation of this compound (Figure 4). Considering transport in a column, Van Genuchten and Dalton [1986] showed that the first order model reproduced the fronts found with the diffusion model well for cylindrical geometry, but less accurately for spherical geometry. Goltz and Roberts [1986] were not able to discriminate between the diffusion and the first order models for their experimental data.

3. Unreacted Shrinking Core Model (USC)

A similar analysis as in Section 2 can be developed for the situation that the original solid particle can not be penetrated by the solute, whereas the solid reaction product has large enough pores to be accessible for the solute. In that case solute diffuses through the stagnant fluid film and product layer that is formed towards the interface where the reaction takes place. The product layer grows in thickness towards the particle interior at the expense of the original solid core. This mechanism, with three resistances to conversion in series, may be described with the unreacted shrinking core (USC) model, Figure 3b.

Other, similar, situations may arise such as when the product layer is inaccessible to the solute (elemental Al oxidation, which stops soon after the product layer is formed) or when the formed product is soluble and poses no resistance for solute mass transfer to the interface (pyrite oxidation). Assuming a first order reaction rate at the product/core interface (that is initially at $R_c = R_0$) the equations are the same as (8) except that (8d) should be replaced by

$$D_s \frac{\partial \bar{c}}{\partial r} = m k_r C_0 \bar{c}_c, \qquad r = R_c, \qquad t > 0 \tag{14a}$$

$$D_s \frac{\partial \bar{c}}{\partial r} = -m C_0 \frac{\partial R_c}{\partial t}, \qquad t > 0 \tag{14b}$$

This set of equations has been used to describe reaction processes in metallurgical and chemical engineering [Levenspiel,1972, Braun et al., 1974]. Exact closed form solutions for this moving boundary problem require a suitable transformation of the variables [Ockendon and Hodgkins, 1975, Davis and Hill, 1982]. The mathematics simplify considerably when the pseudo steady state assumption holds (negligible accumulation of solute in the product layer pores). This assumption is believed valid for the reaction between dilute electrolytes and solid minerals, in view of constraints formulated by Bischoff [1963, 1965], Bowen [1965], Davis and Hill [1982] and Hill [1984]. Assuming pseudo steady state, the generic solution (C) is of the form

$$C = \frac{A(t)}{r} + B(t) \tag{15}$$

This solution in combination with (8c) and (14a) yield two relations between A and B (15), and using (14b) we find for the initial condition $t = 0$, $R_c = R_0$ and a concentration, $c(t)$, in the solution:

$$\int c(t)dt = mR_0C_0\left\{\frac{1}{3}\left[\frac{1}{k_f} - \frac{R_0}{D_s}\right]\left[1 - \left(\frac{R_c}{R_0}\right)^3\right] + (mk_rC_0)^{-1}\left[1 - \frac{R_c}{R_0}\right] + \frac{R_0}{2D_s}\left[1 - \left(\frac{R_c}{R_0}\right)^2\right]\right\} \tag{16}$$

where we may consider the left hand side of (16) as the exposure of the solid phase to the reactant in solution. The time (t_f) for a completed conversion (i.e., $R_c = 0$) when the concentration in solution (c) is constant in time [Levenspiel, 1972] is given by

$$t_f = \frac{\gamma}{c} \tag{17}$$

where

$$\gamma = mR_0C_0\left[\frac{1}{3k_f} + \frac{1}{mk_rC_0} + \frac{R_0}{6D_s}\right] \tag{18}$$

Defining the conversion, X, by

$$X = 1 - \left(\frac{R_c}{R_0}\right)^3 \tag{19}$$

we may rewrite (16) in terms of X. After integration, this yields an expression between time and the degree of conversion. Different functions $P_{i,j}(X)$ are obtained for different particle geometries (j) and rate controlling steps (i), see Levenspiel [1972]. For spherical geometry (j=3), we obtain

$$P_{1,3} = X \tag{20a}$$

$$P_{2,3} = 1 - 3(1 - X)^{2/3} + 2(1 - X) \tag{20b}$$

$$P_{3,3} = 1 - (1 - X)^{1/3}, \tag{20c}$$

for film diffusion (i=1), product layer diffusion (i=2), and reaction rate (i=3) control, respectively. For a constant concentration in the bulk of the solution we obtain from (16)-(20) for the product of the concentration in solution with time:

$$ct = \sum \gamma_{i,j} P_{i,j}(X) \tag{21}$$

It may be worthwhile to note that the expression for film diffusion control (20a) is also found when plate or cylinder geometries are considered, although for those cases different expressions for $\gamma_{i,j}$ result [Levenspiel, 1972, Van der Zee et al., 1989b]. For a plate geometry, a linear relationship between exposure (16) and conversion (X) is found also for reaction rate control. Hence, unless the geometry is well known for the sorbing solid phase, such linearity does not allow any conclusions with regard to the rate controlling mechanism.

When the potential sorption is not well known, it may be uncertain what sorbed amount corresponds to a particular value of X. Therefore discrimination between rate controlling steps may become even more difficult for a small conversion, X, as (16) reduces to a first order expression. This can be shown by Taylor series expansion of the terms containing X in (20b, c) for small X. Then, (16) becomes

$$\int c(t)dt = \frac{mR_0C_0}{3}\left[\frac{1}{k_f} + \frac{1}{mk_rC_0}\right]X, \tag{22}$$

which may be compared with (20a). After rearrangement, this yields

$$X = \frac{3k_fk_r}{R_0(mk_rC_0 + k_f)}\int c(t)dt \tag{23}$$

The integral in (23), in case of constant concentration in the bulk solution, becomes ct. Clearly, the overall reaction rate reduces to an apparent first order reaction rate

$$\frac{dX}{dt} = \bar{k}c(t), \tag{24}$$

where the coefficient $\bar{k}$ equals the lumped coefficient of the integral in (23). Sorption is proportional to X by some stoichiometry factors. For small times, the apparent sorption kinetics are therefore first order irrespective of the rate controlling step, just like (4). Fitting data does not give conclusive evidence in that case with respect to the mechanisms involved, or with respect to their dependency on environmental factors such as the hydrodynamic regime or temperature, which would be needed for extrapolation.

More realistic in practical situations, is the case of a distributed particle size. For an exponential size distribution of spherical particles, the conversion time relationship is given in Appendix 1. The relevance of that exercise is that a unique relationship between ensemble conversion and exposure may be expected for a single as well as for a distributed size. The exposure integral in case of a time dependent concentration reads

$$I = \int c(t) dt. \tag{25}$$

In the case of a constant concentration it becomes simply I = c t.

4. Applicability of HCD and USC Models for Soil

As was shown in the previous sections, it may prove very difficult to discriminate between the very similar first order kinetics, the HCD and the USC models. Therefore it is questionable to ascribe physical meaning to model parameters obtained by fitting the amount of sorbed solute (M, or X multiplied with stoichiometry factors) as a function of time with these models.

In practice, deterministic models, such as the HCD and USC model, have a limited applicability for natural soil systems, as was the case for the first order (chemical) kinetics model. This is due to the poorly defined solid phase matrix. Besides that different sorbing phases may be expected in most soils (different, and sometimes impure, clay and oxide minerals), also the size distributions and geometries may not always be cast into simple mathematical functions, as considered above. Different ways of dealing with this problem were presented in the literature [Van Genuchten and Dalton, 1986, Rasmuson, 1986] for relatively well defined aggregated soils. For soil, sorption rate expressions are often unpredictable, due to the complicated structure and composition of natural soil.

Besides deterministic models that have proven useful in understanding behaviour in soil or soil analogues, it seems therefore worthwhile to develop more global models. Deterministic models, as given above, may help theory development, but have limited applicability to real soil systems. Rather than applying these models to natural soil, in which case the mechanistic background of the description and of the parameters is violated, we can derive simple models, based on a mechanistic description. Rather than simplifying nature, we then simplify the model to account for nature's complexity, where the information requirement is in balance with the information usually available in soil modeling. The understanding obtained with these macroscopic models may help to assess which of the mechanistic descriptions at the microscopic level is realistic.

For the first order and HCD-models it was established (13) that the sorbed amount (M(t)) is given by the concentration (c) in solution multiplied with a function of time. Thus, for these models, the sorption associated with diffusive transfer can be scaled with respect to the concentration in the bulk of the solution. A tenfold increase in c implies a tenfold increase in M(t), see (13). The USC models yield the same, or due to analytical error indistinguishable sorption time curves (usually smooth arcs) as the HCD models. However, the concentration dependence may differ, as assuming pseudo steady state unique relationships between sorption (i.e., conversion) and exposure were derived. For the USC-models the sorbed amount (or conversion, X) is given as a function of c(t) integrated over time! This implies that a tenfold increase in c need not result in a tenfold increase in sorption. Rather, two particles with the same conversion have the same exposure, which may be reached via a large concentration and short contact time or via a small concentration and a large contact time. This difference may be used to assess which of the two models denoted here by HCD and USC is more applicable.

For illustration the conversion of chalcopyrite is shown as a function of the exposure (I) to protons (supplied by H_2SO_4), in Figure 5. This result was obtained using the data of Hiskey and Wadsworth [1974] who described chalcopyrite conversion with the USC-model. Hiskey and Wadsworth [1974] showed that conversion plotted as a function of time for different pH-values does not yield a single curve but a fan of curves. Because in Figure 5 a single curve is found when conversion is plotted as a function of exposure (I = ct at different fixed pH-levels), the USC-model appears more realistic than the first order or the HCD-models. This understanding may be useful to describe the behaviour of pyritic soils found in many coastal plains. Other situations where the USC-model may be valid, are the release of cations by minerals in soil due to the reaction with protons [Van der Zee et al., 1989a]. This reaction is of interest in studies of soil acidification. A third example is the reaction of phosphate (denoted P) with soil. As was shown by Van der Zee et al. [1988] the overall reaction (F) may be described well using the USC-model. For illustration, the measured P-sorption is shown as a function of time for four fixed P-concentrations ranging from 0.03-5 $mol.m^{-3}$ in Figure 6a. In total, 16 experiments with a constant P-concentration and 8 batch experiments with a P-concentration decreasing as a function of time were performed, according to procedures of Van Riemsdijk and Van der Linden [1984] and Van der Zee and Van Riemsdijk [1988]. The sorption rate did not increase a hundred fold, as would agree with the first-order or the HCD-models. Instead, it increased much less and as is shown in Figure 6b, total P-sorption (F), for the indicated concentration range, is described well with a single function of (the logarithm of) exposure. This agrees with USC-theory. Because in the literature many examples can be found, where the first-order or the HCD-models seem appropriate (see also Figure 4), no additional examples are given here, for brevity.

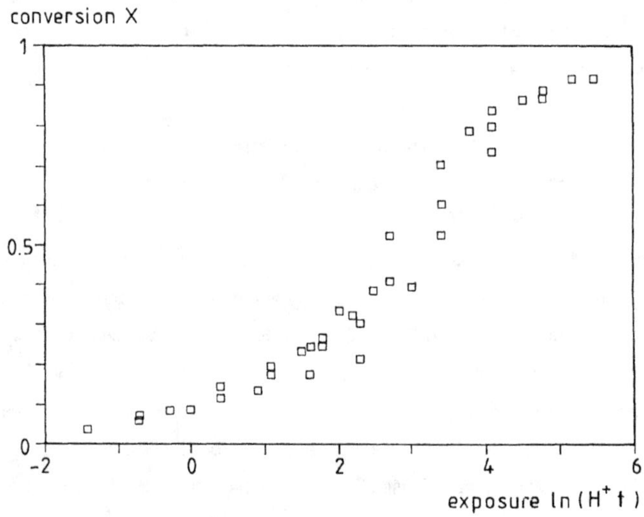

Figure 5: Conversion of chalcopyrite due to exposure to protons (after: Hiskey and Wadsworth [1974]) as a function of log-exposure (ln[H.t)). Concentrations of H_2SO_4 ranged from 0.05 to 2.0 mol/l, and conversion plotted versus time yielded a fan of lines.

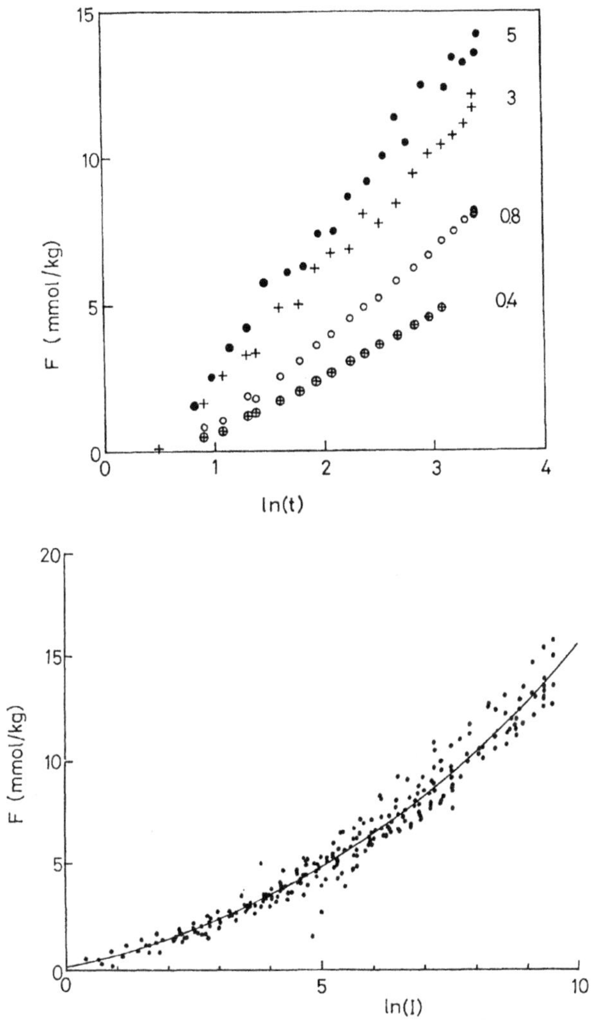

Figure 6: Conversion of metal (Fe, Al) oxides in soil into metal phosphates. Reacted amount of P (F) as a function of time . (Fig. 6a; for four concentrations) and of exposure (Fig. 6b; all data, with line the best fitt.ing second order polynomial).

The above implies, that in order to choose the appropriate microscopic model, it is at least required that (overall) sorption is measured as a function of time and for different concentrations. For a more detailed understanding, it may be necessary to vary other environmental factors, such as temperature, hydrodynamic regime, etc.

TRANSPORT AND KINETICS

Analyses of transport of solutes reacting according to first order kinetics have been given by many soil researchers, in view of the relevance for solutes such as pesticides, nitrate and radionuclides. A number of exact and approximate solutions for irreversible first (and zeroth) order reactions were given by Van Genuchten and Alves [1982]. Decomposition and decay members were also taken into account by Van Genuchten [1985] and Lester et al. [1975]. These rate laws assumed that the overall rate of disappearance or production depended only on the solute concentration. That this assumption is only valid for a limited number of situations was discussed by McLaren [1970] and Parlange et al. [1984]. Barry et al. [1986] discussed an approximation in the case of Michaelis-Menten kinetics. More involved numerical models can be found in the Iskandar [1981].

Solutions for reversible linear first order reaction kinetics were provided by Davidson and McDougal [1973], Hornsby and Davidson [1973], Cameron and Klute [1977] and Rao et al., [1979]. For the nonlinear (Freundlich) sorption case numerical solutions were given by Leistra and Dekkers [1977], and Rao and Jessup [1982] for pesticide sorption, and by Lindstrom et al. [1970] and Weber and Miller [1988] for hydrophobic contaminants. The linear first order kinetics with transport was shown to be mathematically identical to linear sorption with first order mass transfer by Van Genuchten [1981] and Nkedi-Kizza et al. [1984].

Early developments considered the non-equilibrium reactions (or transfer) with purely convective ($D=0$) transport [Thomas, 1944, Klinkenberg, 1948]. Non-equilibrium may give rise to additional dispersion. The same is the case when the flow velocity varies in time [Verhagen, 1978]. Consider the equations

$$\frac{\partial c}{\partial t} = -\left[v\frac{\partial c}{\partial z} + \frac{\rho}{\theta}\frac{\partial q}{\partial t} \right] \tag{26}$$

$$\frac{\partial q}{\partial t} = k_r(k_l c - q) \tag{27}$$

where q is the linearly sorbed concentration (or immobile water concentration). The flow velocity is

$$v = u \qquad nt_p \leq t \leq nt_p + t_f \tag{28a}$$

$$v = 0 \qquad nt_p + t_f \leq t \leq (n+1)t_p \tag{28b}$$

where n is a positive integer, t_p is the period of one flow/stagnation cycle, and t_f is the period of flow. While the flow period, t_f, may be of the order of k_r^{-1}, usually t_p is much larger (e.g., under irrigation conditions). When we assume that $t_f = O(k_r^{-1})$ it can be shown that a reasonable approximation for (27) is obtained by letting changes in q lag behind changes in c [Verhagen, 1978]:

$$q(z,t) = k_l c(z, nt_p) \qquad nt_p < t < nt_p + k_r^{-1} \tag{29a}$$

$$q(z,t) = k_l c(z, t - k_r^{-1}) \qquad nt_p + k_r^{-1} < t < (n+1)t_p \tag{29b}$$

Assuming $t_f < k_r^{-1}$, (26) can be integrated for one cycle (conveniently the first one is taken). Using (29a) we have

$$\frac{\partial c^*}{\partial t}(z,t) + \frac{u}{t_p}\left(1 + k_l\frac{\rho}{\theta}\right)^{-1}\int_0^{t_f}\frac{\partial c}{\partial z}dt = 0 \tag{30}$$

where $c^*(z,t)$ is a continuous function that equals $c(z, nt_p)$ at each time t_p (because $t_p \gg k_r^{-1}$) and may be interpolated rectangularly inbetween. For the first cycle we have $dq(z,t)/dt = 0$ for $t_f < t < t_p$. In (30) we have $c(z,0) = \tilde{c}(z)$, and $c(z,t)$ is given by

$$c(z,t) = \tilde{c}(z - ut) \tag{31}$$

As $\tilde{c}(z-ut) = c^*(z-ut_f, 0)$, also for n>1 we obtain after integration of (30) and using (31):

$$\frac{\partial c^*}{\partial t}(z,t) = \left[t_p\left(1 + \frac{\rho}{\theta}k_l\right)\right]^{-1}[c^*(z-ut_f, nt_p) - c^*(z, nt_p)]; \quad nt_p < t < (n+1)t_p \tag{32}$$

Combination of (32) with the expression found by the Taylor series expansion of $c^*(z-ut_f, nt_p)$ neglecting terms of an order larger than two (valid if vt_f is smaller than a characteristic length over which c varies), we have

$$\frac{\partial c^*}{\partial t} = D\frac{\partial^2 c^*}{\partial z^2} - v_m\frac{\partial c^*}{\partial z}, \tag{33}$$

where

$$v_m = \frac{ut_f}{t_p}\left[1 + \frac{\rho}{\theta}k_l\right]^{-1} = \frac{ut_f}{Rt_p} \tag{34a}$$

$$D = \frac{1}{2}ut_f v_m \tag{34b}$$

In (34a), R is the linear retardation factor. Due to periodic flow and first order kinetics of linear (in c) sorption, we obtain a dispersion with dispersivity equal to

$$L_D^p = \frac{1}{2} u t_f \tag{35}$$

Observe that v_m is the mean front velocity. Expressed more conventionally with respect to the mean water velocity ($\bar{v} = u t_p / t_f$) the dispersivity equals

$$\bar{L}_D = \frac{1}{2} \bar{v} t_p \tag{36}$$

In (36) the parameter t_p represents the frequency with which the flow velocity varies in time. This result indicates that periodic flow conditions may affect the overall dispersive effects significantly for a first order kinetic reaction with approximately the same time constant as flow periodicity. A related and less constrained analysis for sinusoidal solute density oscillations, in the absence of pore scale dispersion and including first order mass transfer into immobile regions, was given by Raats [1973]. In his case of steady state flow, the solute input oscillations gave rise to bypass of the immobile regions. Situations where this may be the case can be encountered in irrigation and flow reactors in industry and waste water renovation.

Because it was realized that apparent first order rate equations for sorption and mass transfer may lump different physical and chemical mechanisms, and have limited validity for different conditions of interest, it was attempted to fill in this equation more mechanistically. Usually this was done on the basis of the HCD-model. Examples were given by Raats [1981, 1984], Van Genuchten and Dalton [1986], Parker and Valocchi [1986], and Rasmuson [1981, 1985] for different geometries. Rasmuson [1985b, 1986] also considered particle size distributions. Of interest is the study by Farrel and Larson [1973] who studied the effect of sinusoidal flow velocity oscillations on transport. Their immobile region was described with the spherical HCD-model and pore scale dispersion in the mobile region was accounted for.

When a soil is aggregated and the first order or HCD-models apply, the non-equilibrium process leads to additional dispersion (see above). For sufficiently large travel distances the front becomes Fickian, which means that the solute fronts, for different depths and times in the Fickian regime, may be scaled onto one curve with the transformation

$$\xi = \frac{z - at}{2\sqrt{(Dt/R)}} \tag{37a}$$

In (37a), the front velocity is a=v/R, with R the linear retardation factor, and D is the diffusion/dispersion coeffient that accounts for "stagnant phase" effects.

The Fickian type of displacement is partly due to the linearity of the reactions assumed. Deviations from such linearity may give rise to completely different behaviour. Besides the differences between the first-order and the HCD-models on one side and the USC-model on the other side, with respect to sorption dependency on concentration (in solution) and time, also the resulting displacement is different. An example is the transport of a solute that reacts according to the USC-model, by Hall et al [1966]. They found, that if product layer diffusion is overall rate controlling for a spherical geometry and pore scale dispersion is neglected, transport is according to a travelling wave. The travelling wave type of displacement implies a limiting $(t \to \infty)$ fixed shape front moving with a fixed velocity through a homogeneous column, at steady state flow. In practise, the limiting front may be reached after short times already. Then, instead of (37a) the fronts for different times and depths can be scaled onto the travelling wave front, with the transformation

$$\eta = z - at \tag{37b}$$

Travelling wave fronts are non-Fickian, because they do not spread with the square root of time. Using the transformation given by (37a) for scaling a travelling wave, would lead to a fan of curves in the concentration ξ-plane, with steeper curves for larger times (instead of time independence of $c(\xi)$).

Also when pore scale dispersion is accounted for, real systems may exhibit a travelling wave type of displacement. This can be illustrated numerically for phosphate transport in a column under field conditions. To describe transport in a column, for phosphate reacting conform the USC-model, the equation

$$\rho \frac{\partial}{\partial t}(F) + \theta \frac{\partial}{\partial t}(c) = \theta D \frac{\partial^2 c}{\partial z^2} - \theta v \frac{\partial c}{\partial z} \tag{38}$$

was solved numerically, for a column with initially zero resident solute. Phosphate adsorption [Van der Zee et al., 1989b] was not considered explicitly. Details for the finite difference solution were provided by Jansen [1985]. Sorption occurring within one second (for small I-values or exposure) was assumed to be instantaneous, while sorption stopped when the polynomial, shown in Figure 6b, reached its maximum (ln(I) = 19). In Figure 7 the front of sorbed solute ($\overline{F} = F / F_m$) is shown for different times. Clearly, when the entire sorption front has entered the soil column it starts to displace in a travelling wave like fashion. For short time periods the front steepens, which is not in agreement with Fickian type of displacement. After some time, the front displaces with time invariant shape and displacement velocity. Thus, because the front width does not increase with travelling distance one could conclude a dispersivity decreasing with z would be appropriate, on the basis of analytical solutions for first order kinetics. Because no analytical solutions are available to describe transport with pore scale dispersion and sorption according to the USC-model, it is uncertain whether the travelling wave type of displacement is due to the USC-model (even for single particle size), follows from a particular size distribution for the soil at hand, or is due to other phenomena leading to highly nonlinear kinetics. The results of Hall et al. [1966] justify the assumption that at least when pore scale dispersion is relatively small, travelling wave type of displacement may be expected. Fitting of their approximation on the front observed in Figure 7 is likely to result in erroneous estimates of model parameters for cases where the particle size is distributed, or mechanisms other than product layer diffusion are rate controlling. Hybrids between the linear first order or HCD-models and the USC model may result in case of non-linear adsorption. Then, non-equilibrium effects may be likewise accounted for in an apparent dispersion coefficient, but a travelling wave may develope [Van der Zee, 1990].

CONCLUSION

During the past decades much use has been made in soil science of overall first order kinetics (with or without additional equilibrium processes) and of diffusion type models. Mechanistic diffusion type models received interest as they provided a better foundation for the apparent rate parameters in first order kinetics models. Two simple mechanistic diffusion models were considered in this paper. For many natural soil systems these models must be considered too simple to be valid. This is due in part to the poorly defined geometry and composition of the sorbing solid phase. For homogeneous aggregates with comparable shapes, approximations involving shape factors may yield good results [Van Genuchten and Dalton, 1985]. For small, microporous particles with large variations in size and shape, it remains questionable whether unique solutions can be obtained from experimental results.

Direct fitting of the diffusion and first order kinetics model on experimental data general by does not allow for a discrimination as to which is the most appropriate model. This is due to data scattering, and to the similarities between sorption-time curves for these models, even for rather well defined systems. Additional complications involved for natural soil are likely to increase the difficulties in assessing the most appropriate model and mechanisms. Hence, such discrimination has to be done on data sets for widely varied conditions. By taking the concentration dependence of sorption into account, we are able to assess for the models considered in this paper, which model may be the most appropriate one. These different concentration dependencies suggest two classes of diffusion type approximations, i.e., sorption depending on the product of concentration with a function of time (HCD) and sorption depending on a function of the product of concentration with time (USC). Only in case hybrids of these models would be more appropriate (e.g. nonlinear intraparticle diffusion) or when other assumptions are violated may these semi-empirical models be unsatisfactory.

The combination of diffusion and first order sorption rates with transport through a column is usually associated with additional front spreading, and thus an increase in the apparent diffusion/dispersion coefficient. Similar to the case of continuous flow [Van Genuchten and Dalton, 1986], periodic flow in combination with first order kinetics leads to additional dispersion, that is proportional with the flow velocity and inversely proportional with the mass transfer or reaction rate parameter. According to the work by Hall et al. [1966], we may expect a travelling wave type of displacement for columns with negligible dispersion when the USC-model (product layer diffusion overall rate controlling) is valid. Then, a steady state front develops that does not spread beyond a certain point. Fitting of a solution of transport with first order kinetics to breakthrough data obtained for different depths, is likely to result in apparent dispersivities decreasing with travel distance. Numerical results obtained for transport and sorption according to a semi-empirical expression developed from the USC-model, also such showed a travelling wave type of displacement. At least for those two cases the limiting apparent dispersivities derived for first order and for (HCD) diffusion affected sorption are not compatible with the observations, as continued front spreading does not occur.

While mechanistic modeling of transport and batch reactors using diffusion models improve our understanding of phenomena in porous media, the application of such analyses to real systems should be done with care. Mostly, by using this approach we implicitly simplify nature into the framework of the model. Then, parameter values obtained have little "mechanistic" meaning, and rate constants reflect the loss rate in the solution rather than the rate of diffusion or of reaction. Predictive possibilities thus remain limited. Instead, we can simplify our models to lumped models without aiming at quantifying physical entities. These models may have better predictive possibilities, while suggesting at the same time, which mechanistic model may be the more appropriate one. Moreover, these lumped models may be more compatible with the information of the "dirty" soil system usually available, and thus be more useful in practice.

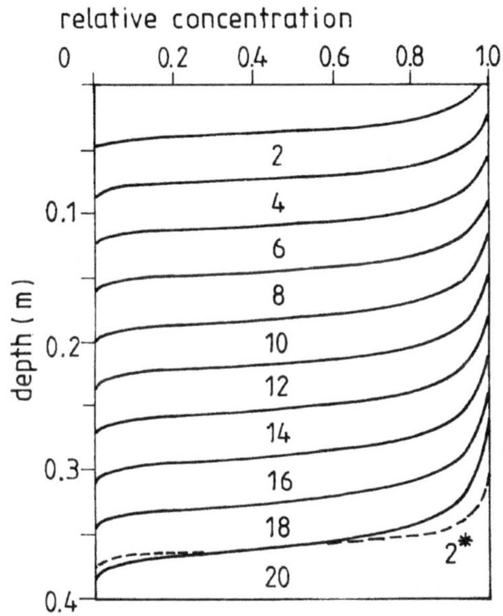

Figure 7: Fronts of relative total sorption ($\bar{F} = F / F_m$) as a function of depth for different times. Sorption was given by the polynomial of F presented in Figure 6. Times in years as indicated at the curves. The curve labelled 2* is for two years shifted to larger depths.

NOTATION

$A(t)$, $B(t)$	functions
c	concentration in solution ($M.L^{-3}$), subscripts A, B, C refer to compounds
c^*	concentration function ($M.L^{-3}$)
$\hat{c}$	concentration in fluid film and particles ($M.L^{-3}$)
$\mathbf{C}$	generic solution of USC-model
C_0	solid phase concentration (1)
D, D_s	diffusion coefficient in column and in particle, respectively ($L^2.T^{-1}$)
E	dimensionless radius
E_1	exponential integral
$\overline{F}$	relative sorbed phosphate amount ($\overline{F} = F / F_m$).
F	overall phosphate sorption ($M.M^{-1}$) with maximum F_m
f	frequency distribution
f_i	function
$\mathbf{G}$	generic solution of (8)
g	function
I	exposure integral (1)
j	mass transfer flux ($M.L^{-3}$), j_A for compound A
k_r, k_f, $\bar{k}$	reaction rate, fluid film mass transfer and apparent first order rate constants, resp. (T^{-1})
k_l	linear sorption coefficient ($L^3.M^{-1}$)
L_D	dispersivity (L)
L_D^p; $\overline{L}_D$	dispersivity expressed as function of maximum and temporal mean velocity, resp. (L).
M	absorbed amount with first order reaction ($M.L^{-3}$)

$\tilde{M}_t$	absorption rate by spherical diffusion without first order reaction ($M.L^{-3}.T^{-1}$)
$\tilde{M}$	absorbed amount in absence of first order reaction ($M.L^{-3}$)
M_t	absorption rat ewith first order reaction ($M.L^{-3}.T^{-1}$)
m	stoichiometric constant (1)
p	functional relationship
P, $P_{i,j}$	function of X (1) with i the rate limiting step, j defining geometry
q	adsorption, ($M.M^{-1}$), constant (Appendix)
R	linear retardation factor (1)
r, R_0, R_c	radial distance, initial particle radius, core radius (L)
r_A	sorption rate ($M.L^{-3}.T^{-1}$)
s	parameter (1)
t, t_p, t_f	time, time of flow cycle, time of completed reaction or of flow period (T)
T	dimensionless time
u	velocity ($L.T^{-1}$) during periods of flow
v	flow velocity ($L.T^{-1}$), parameter (Appendix) (1)
v_m	mean velocity ($L.T^{-1}$)
w_i	function
X, $\overline{X}$	conversion, mean conversion (1)
z	depth (L)
Θ	volumetric water fraction (1)
ρ	dry bulk density ($M.L^{-3}$)
ϵ	particle porosity (1)
ξ	scaled position (1)
γ	lumped parameter ($L^3.M^{-1}.T$)
η	transformed distance (L)

APPENDIX 1: Conversion-time relationship for exponential size distribution

Assuming the USC model valid and one of the mechanisms to be rate controlling, we obtain for constant bulk solution concentration, c,

$$t\frac{c}{\gamma_{i,j}} = P_{i,j}(X) \tag{1.1}$$

Because the functions P may be written as [1-p(1-X)], inversion yields

$$X = 1 - p^{-1}\left(1 - \frac{tc}{\gamma_{i,j}}\right) \tag{1.2}$$

Let the initial size, R_0, be distributed, with size distribution $f(R_0)$, then the mean conversion is

$$\overline{X} = \int_{-\infty}^{\infty} X(R_0) f(R_0) dR_0 \tag{1.3}$$

Expressing time and size with respect to the time of completed conversion and the size of the most reactive particles (with smallest size, $R_0^{(s)}$), gives a dimensionless time (T) and size (E)

$$T = \frac{t}{(\gamma_{i,j}/c)^{(s)}}; \qquad E = \frac{R_0}{R_0^{(s)}} \tag{1.4}$$

where the superscipt (s) refers to the smallest particle size.

We may approximate for two particles with equal conversion the times of completed conversion using

$$\frac{\gamma_{i,j}(1)}{\gamma_{i,j}(s)}=\left[\frac{R_0(1)}{R_0^{(s)}}\right]^{v} \tag{1.5}$$

where the second particle was taken as the reference (smallest) one for convenience, and v is here a parameter equal to 1.5 to 2, 2, and 1 for fluid film, product layer diffusion, and reaction rate control, respectively [Wen, 1968]. Hence, from (1.4) and (1.5) we get

$$\frac{t}{\gamma_{i,j}(1)}=\left[\frac{t}{\gamma_{i,j}(s)}\right]\left[\frac{R_0^{(s)}}{R_0(1)}\right]^{v}=TE^{-v} \tag{1.6}$$

Assuming an exponential size distribution

$$f(R_0)=s\exp[s(R_0^{(s)}-R_0)], \qquad R_0\geq R_0^{(s)} \tag{1.7a}$$

$$f(R_0)=0, \qquad R_0\leq R_0^{(s)} \tag{1.7b}$$

with expectation $R_0^{(s)}+(1/s)$ and variance s^{-2}, we find for (1.3)

$$\overline{X}=1-\int(1-TE^{-v})^{q}s\exp[sR_0^{(s)}(1-E)]R_0^{(s)}dE \tag{1.8}$$

where q follows from the inversion formula of (2) for different i and j, or from an approximation thereof. Integrating only over sizes not yet completely converted at time T, we have two integrals

$$\overline{X} = 1 - \int_{\alpha}^{\infty} (1 - TE^{-v})^{q} s \exp[sR_0^{(s)}(1-E)] R_0^{(s)} dE \tag{1.9a}$$

$$\alpha = 1, \qquad T \leq 1 \tag{1.9b}$$

$$\alpha = T, \qquad T > 1 \tag{1.9c}$$

For the relatively simple case that the reaction rate at the interface of the core is rate controlling (v=1,q=3) this integrates into

$$\overline{X} = -3T^2 sR_0^{(s)} + T^3 sR_0^{(s)}/2 - T^3 (sR_0^{(s)})^2/2$$

$$+ E_1(sR_0^{(s)}) \exp(sR_0^{(s)}) \{3TsR_0^{(s)} + 3(TsR_0^{(s)})^2 + (TsR_0^{(s)})^3/2\}, \qquad T \leq 1 \tag{1.10a}$$

$$\overline{X} = 1 - \exp[sR_0^{(s)}(1-T)] \{1 + 5(TsR_0^{(s)})/2 + (TsR_0^{(s)})^2/2\}$$

$$+ E_1(TsR_0^{(s)}) \exp(sR_0^{(s)}) \{3TsR_0^{(s)} + 3(TsR_0^{(s)})^2 + (TsR_0^{(s)})^3/2\}, \qquad T > 1 \tag{1.10b}$$

These results show that also for a distributed particle size the unique relationship predicted between conversion (for a particle or particle ensemble) and dimensionless time (i.e., the product of concentration times time if c is constant, or the exposure integral, if c varies with time) holds.

REFERENCES

Avnir, D., D. Farin, and P. Pfeifer, Molecular fractal surfaces, Nature, 308, 261-263, 1984

Barry, D.A., J.-Y. Parlange, and J.L. Starr, Interpolation method for solving the transport equation in soil columns with irreversible kinetics, Soil Sci., 142, 296-307, 1986

Bird, R.B., W.E. Stewart, and E.N. Lightfoot, Transport phenomena, Wiley & Sons, New york, 1960

Bischoff, K.B., Accuracy of the pseudo steady state approximation for moving boundary diffusion problems, Chem. Eng. Sci., 18, 711-713, 1963

Bischoff, K.B., Further comments on the pseudo steady state approximation for moving boundary diffusion problems, Chem. Eng. Sci., 20, 783-784, 1965

Bolt, G.H., Movement of solutes in soil: principles of adsorption/exchange chromatography, In G.H. Bolt, (ed.) Soil Chemistry B, Physico-chemical Models, Elsevier, Amsterdam, 285-348, 1982

Bowen, J.R., Comments on the pseudo steady state approximation for moving boundary problems, Chem. Eng. Sci., 20, 712-713, 1965

Braun, R.L., A.E. Lewis, and M.E. Wadsworth, In-place leaching of primary sulfide ores: laboratory leaching data and kinetics model, Metall. Trans. 5, 1717-1726, 1974

Cameron, D.R., and A. Klute, Convective-dispersive solute transport with a combined equilibrium and kinetic adsorption model, Water Resour. Res., 13, 183-188, 1977

Carslaw, H.S., and J.D. Jaeger, Conduction of heat in solids, 2nd ed. Oxford Univ. Press, London, 1959

Crank, J., The mathematics of diffusion, Clarendon Press, Oxford, 1956

Davidson, J.M., and J.R. McDougal, Experimental and predicted movement of three herbicides in water-saturated soil, J. Environ. Qual., 2, 428-433, 1973

Davis, G.B., and J.M. Hill, A moving boundary problem for the sphere, IMA J. Appl. Math. 29, 99-111, 1982

Farrell, D.A., and W.E. Larson, Effect of intra-aggregate diffusion on oscillatory flow dispersion in aggregated media, Water Resour. Res., 9(1), 185-193, 1973

Goltz, M.N., and P.V. Roberts, Interpreting organic solute transport data from a field experiment using physical nonequilibrium models, J. Cont. Hydrol., 1, 77-93, 1986

Hall, K.R., L.C. Eagleton, A. Acrivos, and T. Vermeulen, Pore- and solid diffusion kinetics in fixed bed adsorption under constant pattern conditions, Ind. Eng. Chem. Fundam., 5, 212-223, 1966

Hiemstra, T., J.C.M. de Wit, and W.H. van Riemsdijk, Multi-site proton adsorption modeling at the solid/solution interface of (hydr)oxides: a new approach II: Application to various important (hydr)oxides, J. Colloid Interf. Sci., 133, 105-117, 1989

Hill, J.M., On the pseudo steady state approximation for moving boundary diffusion problems, Chem. Eng. Sci., 39, 187-190, 1984

Hiskey, B.J., and M.E. Wadsworth, Galvanic conversion of chalcopyrite, In: Solution Mining Symposium, F.F. Aplan, W.A. McKinney and A.D. Pernichele (eds.) AIME, New York, 422-445, 1974

Hornsby, A.G., and J.M. Davidson, Solution and adsorbed fluometuron concentration distribution in a water-saturated soil: experimental and predicted evaluation, Soil Sci. Soc. Am. Proc., 37, 823-828, 1973

Iskandar, I.K., (ed.) Modeling wastewater renovation and land treatment, 802 pp., Wiley, New York, 1981

Jaroniec, M., Physical adsorption on heterogeneous solids, Advances in Colloid and Interface Science 18, 149-225, 1983

Klinkenberg, A., Numerical evaluation of equations describing transient heat and mass transfer in packed solids, Ind. Eng. Chem., 40, 1992-1994, 1948

Leistra, M., and W.A. Dekkers, Computed effects of adsorption kinetics on pesticide movement in soils, J. Soil Sci. 28, 340-350, 1977

Lester, D.H., G. Jansen, and H.C. Burkholder, Migration of radionuclide chains through an adsorbing medium, in: Adsorption and ion exchange, Am. Inst. Chem. Eng. Symp. Series, 71, (152), 202-213, 1975

Levenspiel, O., Chemical reaction engineering, 2nd ed. John Wiley & Sons, New York, 1972

Lindstrom, F.T., and L. Boersma, Theory of chemical transport with simultaneous sorption in a water saturated porous medium, Soil Sci., 110, 1-9, 1970

McLaren, A.D., Temporal and vectorial reactions of nitrogen in soil, A review, Can. J. Soil Sci., 50, 97-109, 1970

Miller, C.T., and W.J. Weber, Jr., Sorption of hydrophobic organic pollutants in saturated soil systems, J. Cont. Hydrol., 1, 243-261, 1986

Ockendon, J.R., and W.R. Hodgkins, (eds.) Moving boundary problems in heat flow and diffusion, Clarendon Press, Oxford, 1965

Pannetier, G., and P. Souchay, Chemical kinetics, Elsevier, Amsterdam, 1967

Parker, J.C., and A.J. Valocchi, Constraints on the validity of equilibrium and first-order kinetic transport models in structured soils, Water Resour. Res., 22, 399-407, 1986

Pfeifer, P., D. Avnir, and D. Farin, Ideally irregular surfaces of dimension greater than two, in theory and practise, Surf. Sci., 126, 569-572, 1983

Parlange, J.-Y., J.L. Starr, D.A. Barry, and R.D. Braddock, Some approximate solutions of the transport equation with irreversible reactions, Soil Sci., 137, 434-442, 1984

Nkedi-Kizza, P., P.S.C. Rao, R.E. Jessup, and J.M. Davidson, Ion exchange and diffusive mass transfer during miscible displacement through an aggregated oxisol., Soil Sci. Soc. Am. J., 46, 471-476, 1982

Raats, P.A.C., Propagation of sinusoidal solute density oscillations in the mobile and stagnant phases of a soil, Soil Sci. Soc. Am. Proc., 37(5), 676-680, 1973

Raats, P.A.C., Transport in structured porous media, In: Proc. Euromech. 143, A. Verruit and F.B.J. Barends, (eds.), Balkema, Rotterdam, 221-226, 1981

Raats, P.A.C., Tracing parcels of water and solutes in unsaturated zones, In: B.Yaron, G. Dagan, and J. Goldschmid (eds.) Pollutants in porous media, Ecol. Studies 47, Springer, Berlin, 4-16, 1984

Rao, P.S.C., J.M. Davidson, R.E. Jessup, and H.M. Selim, Evaluation of conceptual models for describing nonequilibrium adsorption-desorption of pesticides during steady flow in soils, Soil Sci. Soc. Am. J., 43, 22-28, 1979

Rao, P.S.C., and R.E. Jessup, Development and verification of simulation models for describing pesticide dynamics in soils, Ecolog. Modeling, 16, 67-75, 1982

Rasmuson, A., Diffusion and sorption in particles and two-dimesional dispersion in a porous medium, Water Resour. Res., 17, 321-328, 1981

Rasmuson, A., The influence of particle shape on the dynamics of fixed beds, Chem. Eng. Sci., 40, 1115-1122, 1985

Rasmuson, A., The effect of particles of variable size, shape and properties on the dynamics of fixed beds, Chem. Eng. Sci., 40, 621-629, 1985b

Rasmuson, A., Modeling of solute transport in aggregated/fractured media including diffusion into the bulk matrix, Geoderma, 38, 41-60, 1986

Rijnaarts, H., and H. Jumelet, Impact of physical/chemical parameters on aerobic biodegradation of alpha-hexachlorocyclohexane (alpha-HCH) in soil and soil suspensions, MSc thesis, Soil Sci. & Plant Nutrition and Microbiology, Agric. Univ. Wageningen, 1987

Thomas, H.C., Heterogeneous ion exchange in a flowing system, Am. Chem. Soc., J. 66, 1664-1666, 1944

Valocchi, A.J., Validity of the local equilibrium assumption for modeling sorbing solute transport through homogeneous soils, Water Resour. Res., 21, 808-820, 1985

Van Genuchten, M.Th., and W.J. Alves, Analytical solutions of the one-dimensional convective-dispersive solute transport equation, USDA Technical Bull. 1661, 149pp., 1982

Van Genuchten, M.Th., Convective-dispersive transport of solutes involved in sequential first-order decay reactions, Computers and Geosciences, 11(2), 129-147, 1985

Van Genuchten, M.Th., and F. N. Dalton, Models for simulating salt movement in aggregated field soils, Geoderma, 38, 165-183, 1986

Van Riemsdijk, W.H., and A.M.A. van der Linden, Phosphate sorption by soils: II., Sorption measurement technique, Soil Sci. Soc. Am. J., 48, 541-544, 1984

Van Riemsdijk, W.H., Reaction mechanisms of phosphate with Al(OH)3 and sandy soil, Dissertation, Agric. Univ. Wageningen, 1979

Van Riemsdijk, W.H., L.K. Koopal, and J.C.M. de Wit, Heterogeneity and electrolyte adsorption: intrinsic and electrostatic effects, Neth. J. Agric. Sci., 35, 241-257, 1987

Van der Zee, S.E.A.T.M., and W.H. van Riemsdijk, Model for long-term phosphate reaction kinetics in soil, J. Environ. Qual. 17, 35-41, 1988

Van Der Zee, S.E.A.T.M., Transport of reactive contaminants in heterogeneous soil systems, Dissertation, Agric. Univ. Wageningen, 1988

Van der Zee, S.E.A.T.M., F. Leus, and M. Louer, Prediction of phosphate transport in small columns with an approximate sorption kinetics model, Water Resour. Res., 25(6), 1353-1365, 1989b

Van der Zee, S.E.A.T.M., W.H. van Riemsdijk, and J.J.M. van Grinsven, Extrapolation and interpolation by time-scaling in systems with diffusion-controlled kinetics and first-order reaction rates, Neth. J. Agricult. Science, 37, 46--60, 1989a

Verhagen, J.H.G., Dispersie van fosfaten in grond, Nota 58, Dept. Hydraulics and Hydrol., Agric. Univ. Wageningen, (in Dutch), 1982

Weber, W.J., Jr., and C.T. Miller, Modeling the sorption of hydrophobic contaminants by aquifer materials, I, Rates and equilibria, Wat.Res., 22, 457-464, 1988

Wen, C.Y., Noncatalytic heterogeneous solid fluid reaction models, Ind. Eng. Chem., 60 (8), 34-54, 1968

Transport in Porous Media **6**: 739–744, 1991
© 1991 *Kluwer Academic Publishers. Printed in the Netherlands.*

A PERTURBATION SOLUTION FOR NONLINEAR SOLUTE TRANSPORT IN POROUS MEDIA

D.O. LOMEN, A.L. ISLAS, X. FAN
Department of Mathematics, University of Arizona, Tucson, Az. 85721 (U.S.A.)

A.W. WARRICK
Department of Soil and Water Science, University of Arizona, Tucson, Az. 85721 (U.S.A.)

ABSTRACT. A perturbation method is applied to the transport equation for a single reactive chemical with nonlinear rate loss relevant for a soil and water system. The results are compared with the linear rate loss case and the effect of different values of the perturbation parameter is shown. Exponential and step sinks modeling water withdrawn from the profile are illustrated.

Key words. Nonlinear solute transport, perturbation method.

1 Introduction

During the last decade the interest in understanding the movement and disposition of pollutants, pesticides, fertilizers and various chemicals in soil water has greatly increased. Theoretical papers using mathematical models are especially useful since many of the physical and chemical constants are not precisely known. Reviews of analytical models of this phenomena using a convection-diffusion partial differential equation have been presented by Nielsen *et al.* (1986) and van Genuchten and Jury (1987). Collections of analytical solutions have been given by van Genuchten and Alves (1982) and Javendel *et al.* (1984). Almost all of the analytical solutions to date have been for linear equations. However nonlinear equations are obtained if more general forms of chemical rate loss are included in the basic transport equation or if a nonlinear equation is used to describe the relation between the solute and absorbed concentrations. In particular Enfield *et al.* (1982) and van der Zee and van Riemsdijk (1987) consider a Freundlich type relation ($s = Kc^n$) where s is the adsorbed concentration and c the solute concentration. Also, van der Zee and Bolt (1988) (as well as many others) consider Langmuir adsorption ($s = \alpha c/(1+\beta c)$) which has a Taylor series expansion as $\alpha c(1 - \beta c + \ldots) = \alpha c - \alpha\beta c^2 + \ldots$. The analysis in this paper includes both a linear term and a quadratic term in the resulting convection-diffusion equation. Bolt (1982), Yates and Enfield (1989) and Bond and Phillips (1989) have discussed analytical and quasi-analytical solutions for movement of chemicals during steady and unsteady flow in unsaturated soils. In this paper we consider a steady water regime for unsaturated flow, with water uptake by plants included. Then the time dependent equations for solute movement (for this particular water flow regime) are solved with a nonlinear rate loss term, a nonlinear isotherm and solute uptake by plant roots included. A perturbation solution is obtained, which relies on the solution of an associated linear problem developed by Lomen *et al.* (1984).

2 The Governing Equations

The equation that describes the transport of a single reactive chemical during steady water flow in a porous medium can be obtained by combining the continuity equation with the apparent convection-diffusion flux density equation. The equation in one dimension is given by

$$\frac{\partial}{\partial t}(\theta(x)c+\rho_b s)=\frac{\partial}{\partial x}\left(D\theta(x)\frac{\partial c}{\partial x}\right)-q(x)\frac{\partial c}{\partial x}-\beta_s\rho_b s-R\theta(x)+\gamma S(x)c \tag{1}$$

where R, a chemical sink term $(moles/cm^3/d)$, and s, the absorbed concentration $(moles/g)$, are both functions of c, the solute concentration $(moles/cm^3)$; θ is the volumetric water content (dimensionless); ρ_b is the bulk density of the porous medium (g/cm^3); D is the apparent diffusion coefficient (cm^2/d); q is the Darcian flux (cm/d); γ is the (dimensionless) "reflection" coefficient associated with solute being taken up by the roots: $\gamma = 0$ is for no reflection; $\gamma = 1$ for total reflection (cf. Hillel 1980, p.246); S is the plant water uptake function (d^{-1}); t is time (d) and x is the distance (cm) from the surface measured positive downwards.

For the linear case, $s = K_d c$ and $R = \beta_1 c$, an analytical solution was found by Lomen *et al.* (1984). In this paper we concentrate on a nonlinear (quadratic) rate loss, assuming R to be of the form

$$R(c)=\beta_1 c+\mu c^2 \tag{2}$$

with $\mu \ll 1$. As in Lomen *et al.* (1984), to simplify the analysis both K_d and D are taken to be zero. Then equation (1) can be written in a simpler form

$$\frac{\partial c}{\partial t}=-V(x)\frac{\partial c}{\partial x}-K(x)c-\mu c^2 \tag{3}$$

where $V(x) = q(x)/\theta(x)$ and $K(x) = \beta_1 + \gamma S(x)/\theta(x)$. Note that $s = K_d c$ could also be used and the result would also be of the same form as equation (3). Also a nonlinear isotherm of the form $s = K_d c + \mu\theta(x)(\beta_s\rho_b)^{-1}c^2$ gives equation (3). (The case for $D \neq 0$ is considered in Lomen, Islas and Warrick (1990)).

In this paper we solve (3) subject to the initial, boundary and limit conditions of the form

$$c(x,0) = 0 \tag{4}$$

$$c(0,t) = f(t) \tag{5}$$

$$\lim_{x\to\infty} c(x,t) = 0 \tag{6}$$

3 The Perturbation Scheme

The assumption that $\mu \ll 1$ allows us to use the method of regular perturbations which assumes the existence of a solution of the nonlinear problem of the form

$$c(x,t)=c_0(x,t)+\mu c_1(x,t)+\mu^2 c_2(x,t)+\ldots \tag{7}$$

where linear equations for $c_0(x,t), c_1(x,t), c_2(x,t), \ldots$, can be obtained. These equations are found by substituting equation (7) into equation (3), arranging it in powers of μ and setting the coefficients of like powers equal to zero (using the fact that μ is arbitrary). If we do this up to order μ^2, we obtain equations for c_0, c_1 and c_2 given by

$$\frac{\partial c_0}{\partial t} = -V(x)\frac{\partial c_0}{\partial x} - K(x)c_0 \tag{8}$$

$$\frac{\partial c_1}{\partial t} = -V(x)\frac{\partial c_1}{\partial x} - K(x)c_1 - c_0^2 \tag{9}$$

$$\frac{\partial c_2}{\partial t} = -V(x)\frac{\partial c_2}{\partial x} - K(x)c_2 - 2c_0c_1 \tag{10}$$

with their corresponding initial and boundary conditions, obtained in the same way from conditions (4) and (5) and given by

$$c_0(x,0) = 0 \qquad c_0(0,t) = f(t) \tag{11}$$

$$c_1(x,0) = 0 \qquad c_1(0,t) = 0 \tag{12}$$

$$c_2(x,0) = 0 \qquad c_2(0,t) = 0. \tag{13}$$

The limit condition (6) is applied to all the $c_i, i = 0, 1, 2, \ldots$ Similar equations can be found for c_3, c_4, ..., and with them a more accurate solution may be obtained. For the purpose of illustration in this paper three terms are sufficient.

The solutions to equations (8)–(13) are readily found with the use of the Laplace Transform; letting $\bar{c}_0$ be the Laplace Transform of c_0 defined by

$$\bar{c}_0(x,s) = \int_0^\infty e^{-st}c_0(x,t)dt \tag{14}$$

If we multiply equation (8) and conditions (11) by e^{-st}, integrate them with respect to t between 0 and ∞, and use definition (14) for the Laplace Transform of c_0, we find that $\bar{c}_0$ satisfies the *ordinary* differential equation

$$\frac{d\bar{c}_0}{dx} = -\frac{s+K(x)}{V(x)}\bar{c}_0 \tag{15}$$

with the initial condition

$$\bar{c}_0(0,s) = F(s) = \int_0^\infty e^{-st}f(t)dt \tag{16}$$

Equation (15) together with condition (16) has the solution

$$\bar{c}_0(x,s) = F(s)e^{-sf_1(x)}e^{-f_2(x)} \tag{17}$$

where

$$f_1(x) = \int_0^x \frac{dy}{V(y)} \tag{18}$$

and

$$f_2(x) = \int_0^x \frac{K(y)}{V(y)}dy \tag{19}$$

Using the properties of the Laplace Transform we can transform (17) back and find that c_0 is given by

$$c_0(x,t) = f(t - f_1(x))H(t - f_1(x))e^{-f_2(x)} \tag{20}$$

where H is the Heaviside function defined as

$$H(y) = \begin{cases} 1, & y > 0 \\ 0, & y < 0 \end{cases} \tag{21}$$

By the same procedure we can solve equation (9) together with conditions (12) for c_1 and equation (10) together with conditions (13) for c_2, and find that to second order we can write c as

$$\begin{aligned} c(x,t) &= c_0 + \mu c_1 + \mu^2 c_2 \\ &= f(t - f_1(x))\, H(t - f_1(x))\, e^{-f_2(x)} \\ &\times \left[1 - \mu f(t - f_1(x)) \int_0^x \frac{e^{-f_2(y)}}{V(y)} dy \right. \\ &+ \left. 2\mu^2 f^2 (t - f_1(x)) \int_0^x frace^{-f_2(y)} V(y) \left(\int_0^y \frac{e^{-f_2(z)}}{V(z)} dz \right) dy. \right] \end{aligned} \tag{22}$$

This solution satisfies condition (6) automatically as shown in the examples. By continuing this process, we will obtain a convergent infinite series for $\mu < 1$.

4 Examples

In this section we illustrate the effect of a second order reaction by comparing the solution listed as equation (22) with that given by Lomen *et al.* (1984). In fact, the first term of the solution (22), that corresponding to c_0, is the solution found by Lomen *et al.* (1984). Here we will show the effect of the first two correction terms, c_1 and c_2, corresponding to μ and μ^2, and the effect of varying μ. We use the Pima clay loam given by Warrick (1975): $K = 9.9e^{0.014h} cm/d$ with h in cm. We'll consider two cases: the exponential and the step sinks. We use the same expressions as in Lomen *et al.* (1984) for $q(x), \theta(x), S(x), f_1(x)$ and $f_2(x)$ and for these cases the integrals occurring in equations (18) and (19) are evaluated analytically. In both examples we take $\theta_{sat} = 0.3$, the reflection coefficient $\gamma = 0.5$, the first order reaction coefficient $\beta_1 = 0.1$ and the input function $f(t) = 1$. In both examples, K_d was taken as zero, the water added at the surface was 1.1 cm/d and 95% of the total water added is assumed to be withdrawn (i.e. the leaching fraction is 0.05).

4.1 *The Exponential Sink*

Consider an exponential sink of the form

$$S(x) = ae^{-bx} \tag{23}$$

where a and b are positive constants. Let l be the leaching fraction, and q_0 the Darcian flux at the surface. Because of continuity, the water removed is a/b given by

$$a/b = q_0(1 - l) \tag{24}$$

so it suffices to specify either a or b along with q_0 and l. Figure 1 shows the effect of adding correction terms: the curve with the largest values of concentration corresponds to c_0; the curve with the smallest values of concentration is $c_0 + \mu c_1$ while the middle curve represents $c_0 + \mu c_1 + \mu^2 c_2$. Here $\mu = 0.01$.

To show the effect of μ we consider the three values 0.01, 0.015 and 0.02 and the solution in equation (22) was compared in Figure 2 with the solution found by Lomen *et al.* (1984). In this figure larger values of μ correspond to curves of smaller values of concentration.

4.2 *The Step Sink*

A double step function was consider here, where 90% of the uptake takes place in the top 20 cm and 10% in the 20-100 cm depth. Figure 3 shows the effect of 1, 2 and 3 terms in the solution given by equation (22) for $\mu = 0.01$. The effect of varying μ is shown in Figure 4 where μ takes the three values 0.01, 0.02 and 0.03.

5 Summary

An asymptotic solution is found to the problem of transport of a chemical with nonlinear (quadratic) rate loss. When a linear rate loss was considered an exact solution was found by Lomen *et al.* (1984). The solution found here is a correction to their solution when a "weak" second order reaction is present, that is, when the coefficient of the quadratic term is much smaller than the linear one. The effect of this second order reaction is studied by its comparison with the linear reaction solution and by its range when the quadratic coefficient is varied. The approach taken in this paper is currently being extended to multidimensional cases.

Acknowledgements

Support for this paper came from the Environmental Protection Agency under Grant No. CR-814243-01-10.

References

Bolt, G.H. (Editor) (1982) *Soil Chemistry. B. Physico-chemical Models.* Amsterdam, Netherlands; Elsevier Science Publishers B.V.

Bond, W.J. & Phillips, I.R. (1989) Cation Transport During Unsteady, Unsaturated Soil Water Flow: A Simple Approximate Solution. *Water Resour. Res.*, in press.

Enfield, C.G., Carsel, R.F., Cohen, S.Z., Phan, T. & Walters, D.M. (1982) Approximating pollutant transport to ground water. *Ground Water*, 20: 711-722.

Hillel, D. (1980) *Fundamentals of Soil Physics.* Academic Press, New York, NY.

Javendel, I., Doughty, C. & Tsang, C.F. (1984) Groundwater Transport: Handbook of Mathematical Models. *Water Resources Monograph, Am. Geophys. Union*, Washington, D.C.

Lomen, D.O., Islas, A.L. & Warrick, A.W. (1990) A Perturbation Solution for Transport and Diffusion of a Single Reactive Chemical with Nonlinear Rate Loss, in Roth, K., Flühler, H., Jury, W.A. & Parker, J.C. (Editors), *Field-Scale Solute and Water Transport Through Soil.* Birkhäuser Verlag, Basel: 281 - 288.

Lomen, D.O., Tonellato, P.J. & Warrick, A.W. (1984) Salt and Water Transport in Unsaturated Soil for Non-conservative Systems. *Agricultural Water Manage.* 8 : 397–409.

Nielsen, D.R., van Genuchten, M.Th, & Biggar, J.W. (1986) Water Flow and Solute Transport Processes in the Unsaturated Zone. *Water Resour. Res.* 22: 895–1085.

van der Zee, S.E.A.T.M., & Bolt, G.H. (1988) Deterministic and Stochastic Modeling of Reactive Solute Transport, in P.J. Wierenga and D. Bachelet (Editors), *Proceedings of Conference on Validation of Flow and Transport Models for the Unsaturated Zone, Res. Rep.* 88-SS-04, N.M. State Univ. Dep. of Agron. and Horticulture, Las Cruces, NM: 435-466.

van der Zee, S.E.A.T.M., & van Riemsdijk, W.H. (1987) Transport of reactive solute in spatially variable soil systems. *Water Resour. Res.*, 23: 2059-2069.

van Genuchten, M.Th. & Alves, W.J. (1982) Analytical Solutions of the One-Dimensional Convective-Dispersive Solute Transport Equation. *Technical Bulletin 1661, U.S. Department of Agriculture, Agricultural Research Service*, Washington, D.C.

van Genuchten, M.Th. & Jury, W.A. (1987) Progress in Unsaturated Flow and Transport Modeling, *Rev. of Geophys.* 25 : 135–140.

Warrick, A.W. (1975) Analytical Solutions to the One-Dimensional Linearized Moisture Flow Equation for Arbitrary Input. *Soil Science* 120: 79–84.

Yates, S.R. & Enfield, C.G. (1989) Transport of Dissolved Substances with Second-Order Reaction. *Water Resour. Res.* 25: 1757–1762.

Transport in Porous Media 6: 745–758, 1991
© 1991 *Kluwer Academic Publishers. Printed in the Netherlands.*

TRACE TYPE FUNCTIONAL DIFFERENTIAL EQUATIONS AND THE IDENTIFICATION OF HYDRAULIC PROPERTIES OF POROUS MEDIA

J. R. CANNON[1]
Department of Mathematics
Lamar University
Beaumont, Texas

PAUL DUCHATEAU[2]
KEN STEUBE[2]
Department of Mathematics
Colorado State University
Fort Collins, Colorado

Summary: The authors give four examples of inverse problems related to fluid flow in a porous media. These are the problems of identifying an unknown source term, and unknown diffusivity, an unknown porosity, and both an unknown diffusivity and porosity. The authors show how these problems can be recast into problems whose equations are well-defined differential equations with trace type functional coefficients. The authors conclude with a discussion of the reformulated problems.

Keywords: Trace Type Functional Differential Equations

MR Classification: 35R.

---

[1]Acknowledgement: supported in part by NSF grant DMS-8901301 and in part by the Texas Advanced Technology Program Grant No. 003581-005.
[2]Acknowledgement: supported in part by ONR contract number N00014-88-K-0224.

1. Introduction

In the context of modeling flow in a porous medium, we are obliged to consider partial differential equations derived from general principles. Such equations apply then not to just one but to a large class of porous media and only become descriptive of a specific porous medium when the ingredients of the partial differential equation (e.g. coefficients, source terms, etc.) are chosen in an appropriate way. In general these ingredients are complicated functions of the independent and dependent variables in the equation, and the problem of determining how they should be chosen in order that the resulting partial differential equation models a given porous medium is very difficult.

We think of the ingredients as characterizing the hydraulic properties of the porous medium under consideration, and the problem of choosing appropriate functions for the ingredients can be viewed as identifying the hydraulic properties of the porous medium. In some cases the necessary hydraulic properties can be determined directly by means of some laboratory procedure, but often an indirect identification procedure must be employed. Typically indirect procedures consist of formulating and solving an appropriate inverse problem. In the most general situation where the ingredients are functions of a mixture of dependent and independent variables, it is not at this point well understood how to formulate (much less solve) an inverse problem appropriate to the identification of the hydraulic properties. However, in certain cases it is an acceptable approximation of reality to suppose that the hydraulic properties are functions of but a single (dependent or independent) variable. In this case it is possible to formulate an inverse problem whose solution will lead to the identification of the desired hydraulic functions. It is our aim here to show how this may be done.

When one begins a study of inverse problems aimed at identifying unknown ingredients in partial differential equations from overspecified data, it soon becomes apparent that there is no well defined procedure for formulating a "well posed" inverse problem. There is considerable flexibility in choosing overspecified conditions and no clear way to tell whether one has chosen the "right" conditions nor if the amount of overspecified data is "enough but not too much" to uniquely determine the desired unknown ingredients. We present here a procedure that may provide insight if not answers to many questions of this nature. We shall show that by reformulating a given inverse problem as a related problem involving a functional partial differential equation, it is possible to see whether the original inverse problem is well formulated. In addition, in situations where it is possible to formulate more than one inverse problem for the identification of given ingredients, this new approach provides a basis for comparing inverse problems in order to see which is "best." Other insights based on this approach will be described below.

The idea of reformulating inverse problems as functional differential equations was initially proposed [1,2,3]. More recent efforts in this direction are described in [4,5,6].

2. Inverse Problems

In order to illustrate our approach we shall consider first several examples of inverse problems that might typically be encountered in the modeling of flow in a porous medium. In the next section we shall show how these inverse problems may be reformulated and in the final section we draw attention to the information that is provided by this reformulation.

Example 2.1 – Identifying an Unknown Source Term

Consider the problem of finding unknown functions $u = u(x,t)$ and $F = F(u)$ that satisfy

$$\begin{array}{ll} u_t(x,t) - u_{xx}(x,t) = F[u(x,t)], & 0 < x < 1, \quad 0 < t < T, \\ u(x,0) = 0, & 0 < x < 1, \\ u(0,t) = f(t), \quad u_x(1,t) = 0, & 0 < t < T, \\ u_x(0,t) = g(t), & 0 < t < T. \end{array} \tag{2.1}$$

Here $f(t)$ and $g(t)$ are assumed to be given data and the unknown source term $F = F(u)$ is supposed to be a function of only the unknown dependent variable $u = u(x,t)$. If $F(u)$ is known a- priori, then specifying both $f(t)$ and $g(t)$ leads to an overdetermined problem. When $F(u)$ is unknown then the problem (2.1) is not overdetermined but note that there is no basis for distinguishing one of the boundary conditions at $x = 0$ as the overspecified condition nor for viewing the other condition then as part of the so-called direct or forward problem. We point out that when two conditions are imposed at a single point (at $x = 0$ in this example) we are to interpret this to mean that in an experimental realization of this mathematical model we force <u>one</u> of the conditions to be satisfied while <u>observing</u> the other. It is not, of course, possible to force both conditions to be satisfied for arbitrary choices of $f(t)$ and $g(t)$.

Note also that while there is no loss in generality in supposing the auxiliary conditions at $t = 0$ and $x = 1$ to be homogeneous, there is also no evident reason to prefer a Neumann condition to Dirichlet condition at $x = 1$. Finally one could ask how the inverse problem is affected if we replace one of the conditions at $x = 0$ with one or both of the following overspecifications:

$$u(x,T) = p(x), \quad 0 < x < 1, \tag{2.2}$$

$$\int_0^1 u(x,t)\,dx = E(t), \quad 0 < t < T, \tag{2.3}$$

In the next sections we shall illustrate an approach that can provide answers to some of these questions.

Example 2.2 – Identifying an Unknown Diffusivity

Consider the problem of finding unknown functions $u = u(x,t)$ and $D = D(u)$ that satisfy

$$\begin{array}{ll} u_t(x,t) = \{D(u)\,u_x(x,t)\}_x, & 0 < x < 1, \quad 0 < t < T, \\ u(x,0) = 0, & 0 < x < 1, \\ u(0,t) = f(t), \quad u_x(1,t) = 0, & 0 < t < T, \\ D[u(0,t)]u_x(0,t) = g(t), & 0 < t < T. \end{array} \tag{2.4}$$

Here $f(t)$ and $g(t)$ are assumed to be given data and the unknown "diffusivity" $D = D(u)$ is supposed to be a function of only the unknown function $u = u(x,t)$. The remarks made in connection with the previous example apply equally well here. In particular, one might consider the affect on the inverse problem if the flux condition at $x = 0$ were replaced by the condition (2.3). In this case, note that the partial differential equation then implies

$$E'(t) = 0 - D[u(0,t)]u_x(0,t),$$

and it is evident that the inverse problem (2.4) is equivalent to the inverse problem in which the flux condition at $x = 0$ is replaced by (2.3). On the other hand, we can show that the condition (2.2) is unsuitable as an overspecified condition in this inverse problem.

Example 2.3 – Identification of a Capacity Term

Consider the problem of finding unknown functions $u = u(x,t)$ and $C = C(u)$ that satisfy

$$\begin{array}{lll} C[u(x,t)]u_t(x,t) = u_{xx}(x,t), & 0 < x < 1, & 0 < t < T, \\ u(x,0) = 0, & 0 < x < 1, & \\ u(0,t) = f(t), \quad u_x(1,t) = 0, & 0 < t < T, & \\ u_x(0,t) = g(t), & 0 < t < T. & \end{array} \tag{2.5}$$

As in the previous examples, $f(t)$ and $g(t)$ are supposed to be given data and the unknown ingredient $C = C(u)$, which may be interpreted as a capacity term, is assumed to depend only on the unknown function $u = u(x,t)$. In each of the examples 2.2 and 2.3, there is just a single unknown ingredient to be found and only one extra condition is specified. Compare these with the following example.

Example 2.4 – Simultaneous Determination of Two Ingredients

Consider the problem of finding unknown functions $u = u(x,t)$ and $K = K(u)$, $C = C(u)$ that satisfy

$$\begin{array}{lll} C[u(x,t)]u_t(x,t) = \{K(u)u_x(x,t)\}_x, & 0 < x < 1, & 0 < t < T, \\ u(x,0) = 0, & 0 < x < 1, & \\ u(0,t) = f(t), & 0 < t < T, & \\ K[u(0,t)]\, u_x(0,t) = g(t), & 0 < t < T, & \\ u(1,t) = h(t), & 0 < t < T, & \\ u_x(1,t) = 0, & 0 < t < T. & \end{array} \tag{2.6}$$

Since we are attempting in this inverse problem to simultaneously determine the unknown capacity term $C = C(u)$ and the conductivity term $K = K(u)$, it seems on an intuitive level that an additional piece of overspecified data ought to be included over what is provided in the previous two examples. At the point $x = 0$ we are already measuring both $f(t)$ and $g(t)$, the Dirichlet and Neumann data consistent with this partial differential equation. In addition we have specified a no flow condition at $x = 1$. However, we shall see in the next section that a simpler inverse problem is obtained if, instead of measuring $u(1,t)$, we specify that

$$\theta[u(1,t)] = h(t), \quad 0 < t < T. \tag{2.7}$$

Here $\theta'(u) \equiv C(u)$ is interpreted physically as moisture content if $C(u)$ represents fluid capacity. As in example 2.2, we see that replacing the flux condition at $x = 0$ with the condition (2.3) leads to an inverse problem that is equivalent to (2.6).

3. Reformulation of Inverse Problems

We show now how each of the examples of the previous section can be reformulated as an initial boundary value problem for an equation of nonclassical form.

Example 3.1 – Identifying an Unknown Source Term

If the function pair $\{u, F\}$ solves the inverse problem (2.1) then it follows that,

$$F[u(0,t)] = f'(t) - u_{xx}(0,t), \quad 0 < t < T.$$

This motivates us to consider the problem of finding an unknown function $u = u(x,t)$ that satisfies,

$$\begin{aligned} &u_t(x,t) - u_{xx}(x,t) = S\{f;u\}(u(x,t)), && 0 < x < 1, \quad 0 < t < T, \\ &u(x,0) = 0, && 0 < x < 1, \\ &u_x(0,t) = g(t), \quad u_x(1,t) = 0, && 0 < t < T, \end{aligned} \tag{3.1}$$

where,

$$\begin{aligned} &S\{f;u\}(z) \equiv f'(p(z)) - u_{xx}(0,p(z)), \\ &p(z) \equiv [u(0,t)]^{-1}(z), \quad 0 < z < u(0,T). \end{aligned} \tag{3.2}$$

We can show that if $u = u(x,t)$ solves the initial boundary value problem (3.1), then the pair of functions $u = u(x,t)$ and $F = S\{f;u\}$ solves the inverse problem (2.1). We have used the condition $u(0,t) = f(t)$ to eliminate the unknown source $F(u)$ from the partial differential equation and replace it with the nonclassical term $S\{f;u\}$. As the notation is intended to indicate, this term depends in a functional way on the overspecified data $f(t)$ and also on the unknown solution $u = u(x,t)$ through the appearance of the terms $u_{xx}(0,t)$ and $[u(0,t)]^{-1}$. Each of these terms involving $u(x,t)$ represents a restriction to the boundary (i.e., the point $x = 0$) of the solution or derivatives of the solution and so we refer to $S\{f;u\}$ as composed of TRACE TYPE FUNCTIONAL terms. Since the trace type functional source $S\{f;u\}$ involves a composition of two trace type functional terms and in addition $S\{f;u\}$ appears in the equation with $u(x,t)$ as its argument, we refer to the equation in (3.1) as a RECURSIVE trace type functional equation. Equations involving trace type functional terms but where no functional contains an argument which depends on the unknown function $u(x,t)$ are called SIMPLE trace type functional equations.

It is evident that in order for the solution of problem (3.1) to go forward it is necessary that $[u(0,t)]$ be invertible. Under appropriate assumptions on the data $g(t)$, it can be shown that the solution of (3.1) satisfies the condition that $u(0,t)$ is invertible. This emphasizes the point that the flux condition in the inverse problem (2.1) is the condition that is controlled while the condition $u(0,t) = f(t)$ is simply observed. In the setting of the problem (3.1) we see that the Dirichlet condition is used to eliminate the unknown source from the equation while the Neumann condition remains as part of the direct problem. All this suggests that in the inverse problem (2.1), the Dirichlet condition should be viewed as the overspecified condition while the Neumann condition is the condition that is taken as part of the direct problem. The significance of this distinction will be discussed in the next section. It is important to notice that since the condition $u(0,t) = f(t)$ is not imposed on the solution of problem (3.1), it is not known a- priori that the function $p(z)$ defined in (3.2) is equal to $f^{-1}(z)$. It can be shown a-posteriori that if $u(x,t)$ solves (3.1) then $p = f^{-1}$; once this fact is established it follows that the pair of functions $u(x,t)$ and $S\{f;u\}$ solve the inverse problem (2.1).

It can be shown that the condition (2.3) cannot be used to eliminate the source $F(u)$ from the equation in (2.1) and we conclude that this condition is not a suitable overspecifcation for the identification of an unknown source. On the other hand, the condition (2.2) can take the place of one of either of the two conditions imposed at $x = 0$. For example, if we replace the flux condition at $x = 0$ in (2.1) with the condition (2.2) then we are led to consider the problem of finding an unknown function $u(x,t)$ that satisfies,

$$\begin{aligned} &u_t(x,t) - u_{xx}(x,t) = \tilde{S}\{f;u\}(u(x,t)), && 0 < x < 1, \quad 0 < t < T, \\ &u(x,0) = 0, && 0 < x < 1, \\ &u(0,t) = f(t), \quad u_x(1,t) = 0, && 0 < t < T, \end{aligned} \tag{3.3}$$

where,

$$\begin{aligned} &\tilde{S}\{f;u\}(z) \equiv u_t(q(z),T) - p''(q(z)), \\ &q(z) \equiv [u(x,T)]^{-1}(z), \quad u(0,T) < z < u(1,T). \end{aligned} \tag{3.4}$$

Remarks similar to those made with respect to the problem (3.1) can be made here. In particular we can ensure that $u(x,T)$ is invertible by suitably controlling the data $f(t)$. In addition, while it is not a-priori, evident that $q = p^{-1}$ this can be established a-posteriori for any solution of (3.3). An important distinction between problems (3.1) and (3.3) however, lies in the fact that (3.3) requires overspecified data measured on the interior of the spatial domain $\{0 < x < 1\}$ while the data for (3.1) is measured on the (spatial) boundary at $x = 0$. Thus, while the functional forms that appear in (3.4) do involve trace operators acting on $u(x,t)$, these are traces on the "temporal boundary," i.e., they involve restrictions to $t = T$. Experimentally it is generally easier to obtain measurements on the spatial boundary than in the interior of the spatial domain.

Example 3.2 – Identifying an Unknown Diffusivity

If the pair of functions $\{u, D\}$ solves the inverse problem (2.4), then it follows that

$$D[u(0,t)]\,u_x(0,t) = g(t), \quad 0 < t < T,$$

and we are led to consider the problem of finding a function $u(x,t)$ that satisfies,

$$\begin{aligned} &u_t(x,t) = \{D[g;u](u)\,u_x(x,t)\}_x, && 0 < x < 1,\ \ 0 < t < T, \\ &u(x,0) = 0, && 0 < x < 1, \\ &u(0,t) = f(t), \quad u_x(1,t) = 0, && 0 < t < T, \end{aligned} \tag{3.5}$$

where,

$$D[g;u](z) \equiv \frac{g[f^{-1}(z)]}{u_x(0, f^{-1}(z))}, \quad 0 < z < f(T). \tag{3.6}$$

Notice that in problem (3.5) the condition $u(0,t) = f(t)$ is imposed on the solution so that we are entitled to use f^{-1} for the inverse of $u(0,\cdot)$ and of course $f(t)$ can be controlled so as to ensure that this inverse exists. The overspecified flux condition in the inverse problem has been used to eliminate the unknown diffusivity, and it is clear from (3.6) that if $u(x,t)$ solves (3.5), then the pair $\{u, D[g;u]\}$ solves the inverse problem (2.4).

Evidently in this inverse problem it is the flux condition at $x = 0$ that plays the role of the overspecified (observed) condition while the Dirichlet condition at $x = 0$ remains as part of the direct problem (it is the controlled condition). It is not hard to see that the condition (2.2) does not permit the elimination of the unknown diffusivity so that this condition is unsuitable as an overspecification in the identification of an unknown diffusivity. The condition (2.3) has been shown to be equivalent to the flux condition for purposes of identifying an unknown diffusivity. Although the inverse problems are equivalent, the measurement required for condition (2.3) would generally be experimentally easier to perform than a flux measurement.

Example 3.3 – Identification of a Capacity Term

If the function pair $\{u, C\}$ solves the inverse problem (2.5), then it follows that,

$$C[u(0,t)]f'(t) = u_{xx}(0,t), \quad 0 < t < T.$$

Then we are motivated to consider the problem of finding an unknown function $u = u(x,t)$ that satisfies,

$$\begin{aligned} &C[f;u]u_t(x,t) = u_{xx}(x,t), && 0 < x < 1, \ \ 0 < t < T, \\ &u(x,0) = 0, && 0 < x < 1, \\ &u_x(0,t) = g(t), \quad u_x(1,t) = 0, && 0 < t < T, \end{aligned} \tag{3.7}$$

where,

$$\begin{aligned} &C[f;u](z) \equiv \tfrac{u_{xx}(0,p(z))}{f'[p(z)]} && 0 < z < u(0,T), \\ &p(z) \equiv [u(0,\cdot)]^{-1}(z), && 0 < z < u(0,T). \end{aligned} \tag{3.8}$$

The remarks made in connection with example 3.1 can also be made here.

Example 3.4 – Simultaneous Determination of Two Ingredients

If the triple of functions $\{u, C, K\}$ solves the inverse problem (2.6), then

$$K[u(0,t)]\,u_x(0,t) = g(t), \quad 0 < t < T,$$

and

$$C[u(1,t)]h'(t) = \{K(u)u_x\}_x \text{ at } x = 1, \quad 0 < t < T.$$

This second expression does not lead to a <u>simple</u> elimination of $C(u)$ from the partial differential equation. At best we may be able to express C in terms of a trace type functional conductivity and other trace type terms. Apparently the measurement of u at the boundary permits identification of C in example 3.3, the case of constant conductivity, but this overspecification leads to a more complicated problem when the conductivity depends on u. However, suppose the condition $u(1,t) = h(t)$ is replaced by the condition (2.7) where θ denotes an antiderivative for C. If $\{u, C, K\}$ is a solution of (2.6) we have

$$\theta'[u(1,t)]u_t(1,t) = h'(t), \quad 0 < t < T.$$

Then we are motivated to consider the problem of finding an unknown function $u = u(x,t)$ solving

$$\begin{aligned} &C[h;u]u_t(x,t) = \{K[g;u]\,u_x(x,t)\}_x, && 0 < x < 1, \quad 0 < t < T, \\ &u(x,0) = 0, && 0 < x < 1, \\ &u(0,t) = f(t), && 0 < t < T, \\ &u_x(1,t) = 0, && 0 < t < T. \end{aligned} \tag{3.9}$$

where,

$$\begin{aligned} &C[h;u](z) \equiv \tfrac{h'(q(z))}{u_t(1,q(z))} \\ &q(z) \equiv [u(1,\cdot)]^{-1}(z), \quad 0 < z < u(1,T). \end{aligned} \tag{3.10}$$

We can show (with some difficulty) that it is possible to control the data function $f(t)$ so that both f and $u(1,t)$ are invertible. This ensures that the solution of (3.9) can go forward.

If $u(x,t)$ solves (3.9) then the flux condition at $x = 0$ appearing in the inverse problem is automatically satisfied for conductivity function $K = K[g;u]$. Similarly we can show that the condition (2.7) is satisfied for θ an antiderivative of $C[h;u]$. Thus a solution of the trace type functional problem (3.9) leads to a solution of the associated inverse problem.

Note that in this problem two overspecifications were required in order to eliminate the two unknown coefficients C and K. Moreover, the fact that the partial differential equation is second order in x requires one boundary condition at each of the ends $x = 0$ and $x = 1$ in order for the direct problem to be well posed. This forces the overspecified conditions to likewise be placed one at each end of the interval $(0,1)$ if these conditions are to be measured on the boundary. Since neither of the conditions (2.2), (2.3) turns out to be a suitable overspecification for this inverse problem, it appears that <u>boundary</u> overspecifications are necessary here.

4. Observations Based on the Reformulation

The examples considered here involve inverse problems in which the unknown ingredients are functions of the unknown solution $u = u(x,t)$ only. Problems in which the unknown ingredients are functions of more than a single variable are more difficult and may be the subject of future studies. Inverse problems in which the unknown ingredients depend on a single independent variable have been considered in [6]. Such problems, when reformulated as trace type functional problems, lead invariably to trace type functional equations of SIMPLE type; i.e., no trace type functional term contains any argument which depends on the unknown solution u. In particular, this precludes the composition of two or more trace type functional terms.

It is well known that unknown ingredient inverse problems of the type considered here are very unstable. The source of this instability lies in the fact that the "coefficient to data mapping" which associates overspecified data values as output corresponding to coefficient input is a "smoothing operator." That is, variations in the coefficients of the partial differential equation are smoothed out by the solution operator of the PDE and produce limited response in the overspecified data; varying the input to the coefficient to data mapping has limited effect on the output of the mapping. In solving an inverse problem, we are attempting to invert this mapping and it follows that small errors in the overspecified data are magnified to produce large variations in the determined coefficients. The reformulation of the inverse problem as a trace type functional problem provides a more detailed view of this instability.

For example, when the inverse problem (2.1) is reformulated as the trace type functional problem (3.1), we obtain what appears to be a classical initial boundary value problem for a forced heat equation in which the existence and regularity of the solution $u(x,t)$ is controlled by the smoothness properties of the source term $S[f;u]$. Appearance of the expression $u_{xx}(0,p(z))$ in (3.2) causes the source term to be extremely sensitive to variations in $u(x,t)$ and, of course, the solution $u(x,t)$ is going to depend in turn on the source. It is this recursive dependence of the solution $u(x,t)$ on itself that is the source of the instability in the inverse problem.

By comparing (3.4) with (3.2) we see that the alternative reformulation (3.3) for the inverse problem (2.1) offers no appreciable improvement in the instability involved in identifying an unknown source. On the other hand, when the inverse problem (2.4) for an unknown diffusivity is reformulated as (3.5), it seems from (3.6) that a less unstable problem is obtained. This conclusion is based on the observation that the trace type functional "recipe" (3.6) involves lower order derivatives of the solution $u(x,t)$ than those appearing in the recipes (3.2) and (3.4). (Recall that for the heat equation one time derivative is equivalent in some sense to two space derivatives). We summarize these remarks by saying that source identification problem seems to be more unstable than the problem of identifying an unknown diffusivity coefficient.

In connection with the inverse problem (2.6) for the simultaneous determination of two unknown coefficients it was remarked that one feels intuitively that at least two pieces of overspecified data are needed. In [7] Hornung uses a numerical experiment for a specific example to show that one condition is almost certainly insufficient and that two conditions may be enough to identify two unknown coefficients. When (2.6) is reformulated as (3.9) it is clearly seen that a single overspecified condition is insufficient to eliminate both unknown coefficients. On the other hand, two overspecifications allow both unknown coefficients to be eliminated leaving a classical form initial boundary value problem. Although the partial differential equation in the problem (3.9) contains nonclassical coefficients, we can use more or less standard approaches to prove the existence of a unique solution to the problem. This solution leads in turn to a solution for the original inverse problem (2.6). Moreover, it is evident that the overspecified condition (2.7) leads to a less complicated trace type functional problem than is obtained when $u(1,t)$ is specified. This provides the basis for the statement that a more stable inverse problem is obtained if we measure $\theta[u(1,t)]$ instead of $u(1,t)$ as part of the overspecified data.

In connection with this same example we made the remark that one boundary condition is needed at each end of the interval $(0,1)$ in order for the direct problem to be well posed and that this forces us to place one overspecified condition at each end of the interval. One observes that if we were to consider this inverse problem on the semi-infinite interval $(0,\infty)$, there are not enough boundary positions to permit us to measure sufficient overspecified boundary data to allow simultaneous determination of the two unknown coefficients. This is a generalization of a similar observation made in [8] in connection with simultaneous determination of two unknown constant coefficients. Finally we remark that measuring three (appropriate) pieces of overspecified data seems to permit simultaneous identification of C, K, and F although the problem obtained is extremely unstable.

In each of the inverse problems considered here we have seen that when two conditions are imposed at a single point on the boundary then one of the conditions can be used to eliminate an unknown ingredient while the other remains as part of the direct problem obtained when the inverse problem is reformulated. Note also that there is no flexibility in choosing which condition is which; invariably only one of the conditions is suitable for eliminating an unknown ingredient so as to obtain a trace type functional problem. This suggests that the condition used to eliminated the unknown ingredient is most directly linked with that unknown ingredient and hence ought to be distinguished as the overspecified (observed) condition in the inverse problem.

This distinction may have relevance for so-called output least squares approaches to inverse problems. In this approach, a subset of the auxiliary conditions in the inverse problem is selected (somewhat arbitrarily) so as to have a well posed direct problem. The remaining conditions are used to build an objective function which is then optimized over the set of admissable unknown ingredients. Our observations suggest that the choice of which conditions are to be used in the objective function ought not to be made arbitrarily. Instead, those conditions which our approach identifies as the overspecified conditions should be used in the objective function since these are the conditions on which the unknown ingredient seems to depend most directly.

Finally, we point out that in the inverse problem (2.1), (2.4) and (2.5), we have a choice of boundary conditions that may be imposed at $x = 1$. By selecting various choices of condition and analyzing the associated trace type functional problem, it is possible to examine the subtle effects of the nonessential conditions on these inverse problems.

References

1. Cannon, J. R. and DuChateau, P., An inverse problem for an unknown source term in a heat equations, JMAA, vol. 75(2), 1980.
2. Cannon, J. R. and DuChateau, P., Weak solutions $u(x,t)$ to a partial differential equation with coefficients that depend on $u[y, \phi_j(t, u(x,t))]$ $j = 1, \ldots, k$, JDE, vol. 42(3), 1981.
3. Cannon, J. R. and DuChateau, P., An inverse problem for an unknown source term in a wave equation, SIAM J Appl Math, vol. 45(3), 1983.
4. Cannon, J. R. and Yin, Hong-ming, A class of nonlinear, nonclassical parabolic equations, JDE, vol. 78, 1989.
5. Cannon, J. R. and Yin, Hong-ming, A class of multidimensional nonclassical parabolic problems, Differential Eq and Applic., 1989.
6. Cannon, J. R., DuChateau, P., and Steube, K., Inverse problems and trace type functional differential equations, (preprint).
7. Hornung, U., Identification of nonlinear soil physical parameters from an input-output experiment, Progress in Scientific Computing, Birkhauser/Boston, 1983.
8. Cannon, J. R. and DuChateau, P., Determination of unknown physical properties in heat conduction problems, Int J Eng Sci, vol. 11, 1973.

Transport in Porous Media 6: 759–769, 1991
© 1991 Kluwer Academic Publishers. Printed in the Netherlands.

Parameter Identification in a Soil with Constant Diffusivity

DAVID ZACHMANN
Department of Mathematics, Colorado State University
Fort Collins CO 80526 USA
and
IAN WHITE
Division of Environmental Mechanics, CSIRO, Canberra Australia

Abstract. We consider infiltration into a soil that is assumed to have hydraulic conductivity of the form $K = K_s e^{\alpha h}$ and water content of the form $\theta = \beta K - \theta_r$. Here h denotes capillary pressure head while K_s, α, β and θ_r represent soil specific parameters. These assumptions linearize the flow equation and permit a closed form solution that displays the roles of all the parameters appearing in the hydraulic function K and θ. We assume K_s and θ_r to be known. A measurement of diffusivity fixes the product of α and β resulting in a parameter identification problem for one parameter. We show that this parameter identification problem, in some cases, has a unique solution. We also show that, in some cases, this parameter identification problem can have multiple solutions, or no solution. In addition it is shown that solutions to the parameter identification problem can be very sensitive to small changes in the problem data.

KEY WORDS. Parameter identification problem, constant diffusivity, Darcy's law.

1. Introduction

Darcy's law and conservation of fluid volume combine to yield the diffusivity form of the moisture flow equation,

$$\frac{\partial \theta}{\partial t} = \nabla \cdot \left(D \nabla \theta \right) - \frac{\partial K}{\partial z} \tag{1.1}$$

where θ represents the volumetric moisture content, K the unsaturated hydraulic conductivity and z the vertical, downward coordinate. If $\theta = \theta(h)$ and $K = K(h)$ are assumed to be single valued functions of capillary pressure h, then the diffusivity D is defined by

$$D = K(h)/\theta'(h) \tag{1.2}$$

Using the θ-h and K-h relationships, D and K can be expressed as a function of θ. The units of h, z, K and D are respectively, [L], [L], [L/T] and $[L^2/T]$.

In general, equation (1.1) is a nonlinear, parabolic, convection-diffusion equation which is not solvable by analytical means. Equation (1.1) can be linearized and, in many cases, solved exactly if K and θ are assumed to have the forms

$$K = K_s\, e^{\alpha h} \tag{1.3}$$

$$\theta = \beta K + \theta_r \tag{1.4}$$

where K_s, α, β and θ_r are constants for a given porous medium. Under assumptions (1.3) and (1.4), the diffusivity, D, defined by equation (1.2) is found to be constant,

$$D = (\alpha\beta)^{-1},$$

and the flow equation (1.1) simplifies to the constant coefficient, linear equation

$$\alpha\beta\frac{\partial\theta}{\partial t} = \nabla^2\theta - \alpha\,\frac{\partial\theta}{\partial z} \tag{1.5}$$

The application of the conductivity relationship (1.3) in the analysis of steady flow has been used extensively by a number of researchers. See for example Philip(1968, 1969, 1972), Lomen and Warrick(1974), Raats(1970) and Zachmann and Thomas(1973). Assumptions (1.3) and (1.4) have been used to model transient, partially saturated flow. See Warrick(1974).

The approximations (1.3)-(1.4) to real soils are most closely satisfied from the midrange of moisture contents upward to saturation, $\theta = \theta_s$. The constant θ_r is often taken to be the value of θ at the dry end of the range on which the approximations (1.3) and (1.4) are used. For equations (1.3) and (1.4) to hold near saturation, then K_s is the saturated conductivity and β is related to K_s by

$$\beta = (\theta_s - \theta_r)/K_s \tag{1.6}$$

In applications, the parameters θ_s and θ_r are usually fairly easy to estimate. If θ_s and θ_r are known, or have been estimated, then the porous medium under consideration can be characterized if the only the parameters α and β can be found. Once β is known, equation (1.6) can be used to determine K_s.

One of the goals of this paper is to show how simple, well designed flow experiments can be used to uniquely determine the parameters α and β. A second aim of this paper, is to demonstrate that, by simply collecting data at inappropriate times, one can easily produce an ill-posed problem for the determination of the parameters α and β.

2. Parameter Identification

We assume that all of the flow parameters are known except for α and β. Our first step in identifying α and β from data collected in an infiltration experiment is to determine the diffusivity $D = (\alpha\beta)^{-1}$. The determination of D for constant flux absorption into a material with constant D is considered by White *et al.* (1979). Here we only point out that a closed form for the flux concentration relation facilitates the determination of D in this case. For details on the procedure for fixing a D-value from an infiltration experiment see the paper of White *et al.* (1979).

Since $D = (\alpha\beta)^{-1}$, the diffusivity value fixes the product of α and β. Thus, at this point the parameter identification problem reduces to finding just one constant, either α or β.

Now to find the one remaining unknown parameter, say α, consider a one dimensional infiltration experiment governed by

$$\alpha\beta\frac{\partial\theta}{\partial t} = \frac{\partial^2\theta}{\partial z^2} - \alpha\frac{\partial\theta}{\partial z} \tag{2.1}$$

$$\theta(z,0) = \theta_r \tag{2.2}$$

$$-D\frac{\partial\theta}{\partial z} + K = Q \tag{2.3}$$

where the initial water content θ_r is constant and the constant surface flux satisfies $Q < K_s$. If all the coefficients are known in equations (2.1)-(2.3), then the initial boundary value problem (2.1)-(2.3) is well posed. In the case under consideration, α is considered to be unknown. In an effort to determine the parameter α we assume that at some time t^* the water content at the surface, $\theta^* = \theta(0,t^*)$, is measured. Then, we consider the parameter identification problem consisting of equations (2.1)-(2.3) and the "extra data", or "overspecification" represented by θ^* for the determination of α. Parameter identification problems similar to this have been considered by many researchers including, Cannon and Zachamnn(1982), Hornung(1983), Kool *et al.*(1985), Milly(1987), Perroux *et al.*(1982), Poulovassilis *et al.*(1988), Towner(1982) and Zachmann *et al.*(1981,1982).

The actual analysis of the parameter identification problem described in the last paragraph will be carried out in terms of dimensionless variables. These dimensionless variables permit a more compact presentation of some of the algebraic manipulations required to settle the parameter identification problem. However, the introduction of dimensionless variables may obscure the simple argument that drives the analysis of the parameter identification problem. In an effort to make the following analysis more understandable, we begin with an outline of the parameter identification analysis. We present this outline in terms of physical, rather than dimensionless, variables.

The first step in describing the parameter identification procedure is to point out that, if all flow parameters except α and β, are known, then the solution of equations (2.1)-(2.3) has the form

$$\theta(z,t) = F(z,t;\alpha,\beta).$$

The last equation is meant to indicate that θ is a function of z an t that depends on the choice of the parameters α and β. Measuring θ at $z = 0$ time t^* yields an equation involving the parameters α and β. Namely,

$$F(0,t^*;\alpha,\beta) = \theta^*.$$

The last equation can be viewed as requiring that (α,β) lie on some curve C_1 in the $\alpha\beta$-plane. Recall that we assumed that a diffusivity measurement has fixed the value of the diffusivity D. Thus, $\alpha\beta = D^{-1}$ defines a second curve C_2 in the $\alpha\beta$-plane. From a geometric point of view, the unique solvability of the parameter identification problem reduces to asking whether the curves C_1 and C_2 intersect in exactly one point. To carry out an algebraic analysis of the parameter identification problem we consider

$$F(0,t^*;\alpha,D^{-1}/\alpha) = \theta^*$$

and ask whether there exists a unique α satisfying this equation. We now turn to an algebraic analysis of the parameter identification problem using dimensionless variables.

In terms of the dimensionless variables

$$T = \frac{\alpha t}{4\beta} \qquad \text{and} \qquad Z = \frac{\alpha z}{2},$$

the solution to the initial-boundary value problem (2.1)-(2.3) has been shown by Warrick(1974) to be

$$\begin{aligned}\theta(Z,T) = \theta_r + \beta Q\Bigg\{ & \frac{1}{2}\,\text{erfc}\left(\frac{Z}{2\sqrt{T}} - \sqrt{T}\right) \\ & + 2\sqrt{T/\pi}\,\exp\left[-\left(\frac{Z}{2\sqrt{T}} - \sqrt{T}\right)^2\right] \\ & - (Z + 2T + \tfrac{1}{2})\exp(2Z)\,\text{erfc}\left(\frac{Z}{2\sqrt{T}} - \sqrt{T}\right)\Bigg\}\end{aligned}$$

so the surface moisture content is given by

$$\theta(0,T) = \theta_r + \beta Q\left\{\frac{1}{2}\,\text{erfc}\,(-\sqrt{T}\,) + 2\sqrt{T/\pi}\,e^{-T} - (2T + \tfrac{1}{2})\text{erfc}(\sqrt{T})\right\} \tag{2.4}$$

If we let $\gamma = \alpha\sqrt{(Dt/4)}$ and use $D = (\alpha\beta)^{-1}$, then we find

$$\gamma^2 = \alpha^2\frac{t}{4\alpha\beta} = T.$$

Therefore, with θ_o defined by $\theta_o = \theta(0,t)$,

$$\frac{\theta_o - \theta_r}{\beta Q} = \alpha\frac{\theta_o - \theta_r}{D^{-1}Q} = \alpha\sqrt{\frac{Dt}{4}}\sqrt{\frac{D}{t}}\,\frac{2}{Q}\,(\theta_o - \theta_r) = \gamma\,\frac{2}{Q}\sqrt{\frac{D}{t}}\,(\theta_o - \theta_r)$$

so (2.4) can be rewritten in terms of γ as

$$\frac{2}{Q}\sqrt{\frac{D}{t}}\,(\theta-\theta_r)\,\gamma = \left[\frac{1}{2}\,\mathrm{erfc}(-\gamma) + \frac{2}{\sqrt{\pi}}\,\gamma\, e^{-\gamma^2} - 2(\gamma^2+\tfrac{1}{2})\mathrm{erfc}(\gamma)\right] \tag{2.5}$$

Using $\mathrm{erfc}(\gamma) = 1 - \mathrm{erf}(\gamma)$ and $\mathrm{erf}(-\gamma) = -\mathrm{erf}(\gamma)$ allows (2.5) to be simplified to read

$$\frac{2}{Q}\sqrt{\frac{D}{t}}\,(\theta-\theta_r)\,\gamma = F(\gamma) \tag{2.6}$$

where

$$F(\gamma) = (1+2\gamma^2)\mathrm{erf}(\gamma) + \frac{2}{\sqrt{\pi}}\,\gamma\, e^{-\gamma^2} - 2\gamma^2 \tag{2.7}$$

The identification of α and β from a measurement of water content on the surface proceeds as follows. Suppose that at some time $t = t^*$ the surface water content is measured to be $\theta_o = \theta^*$. Then (2.6) will uniquely identify γ if the line

$$Y = m\gamma \qquad , \; m = \frac{2}{Q}\sqrt{\frac{D}{t^*}}\,(\theta^* - \theta_r)$$

crosses the curve

$$Y = F(\gamma)$$

in exactly one place. Once γ is known, then α can be recovered from

$$\alpha = \frac{2\gamma}{\sqrt{Dt^*}} \qquad \text{and } \beta \text{ from} \qquad \beta = \frac{1}{\alpha D}.$$

To determine if (2.6) is uniquely solvable for γ, note that $F(\gamma)$ satisfies $F(0) = 0$ and $\lim_{\gamma\to\infty} F(\gamma) = 1$. Also,

$$F'(\gamma) = \frac{4}{\sqrt{\pi}}\, e^{-\gamma^2} + 4\,\gamma\,\mathrm{erf}(\gamma) - 4\gamma$$

and

$$F''(\gamma) = -\,4\,\mathrm{erfc}(\gamma)$$

so that the graph of $Y = F(\gamma)$

- is concave down
- has slope $\frac{4}{\sqrt{\pi}}$ at $\gamma = 0$

- is asymptotic to $Y = 1$ as $\gamma \to \infty$.

From these properties of F, it follows that the line $Y = m\gamma$ intersects the curve $Y = F(\gamma)$ exactly once if and only if m is less than $\frac{4}{\sqrt{\pi}}$. The condition for the unique solvability of (2.6) is

$$t^* > \frac{\pi D}{4Q} (\theta^* - \theta_r) \tag{2.8}$$

Since D is a constant and $(\theta^* - \theta_r)$ is bounded above by $(\theta_s - \theta_r)$, it is clear that the solvability condition (2.8) can be met by choosing t^* sufficiently large. This development shows that α and β are uniquely determined *if the measurement of water content is made at a sufficiently large time.* As Figure 1a, with $m \simeq 4$, shows, if the measurement of water content is made too early, then the inverse problem whose kernel lies in equation (2.6) is ill-posed in the sense that no solution exists. Figure 1b, with $m \simeq 1.5$, shows that a sufficiently large choice of t^* leads to a well posed parameter identification problem. Figure 1b also shows that even, in the case $m > \frac{4}{\sqrt{\pi}}$, when the parameter identification problem has unique solution, that solution can be very sensitive to small changes in the problem data. To see this, consider the limiting case in which m approaches $\frac{4}{\sqrt{\pi}}$ from above.

Next consider the possibility of determining α and β from a measured capillary pressure h^* at time t^*. In this case, it follows from (1.4) that

$$\theta_o - \theta_r = \beta K = \beta K_s e^{\alpha h^*} = (\theta_s - \theta_r)\exp\left(\frac{2\gamma}{\sqrt{Dt^*}} h^* \right)$$

so equation (2.6) reads

$$\frac{2}{Q}\sqrt{\frac{D}{t^*}} (\theta_s - \theta_r)\, \gamma \exp\left(\frac{2\gamma}{\sqrt{Dt^*}} h^* \right) = F(\gamma) \tag{2.9}$$

or, more compactly,

$$G(\gamma) = F(\gamma) \tag{2.10}$$

where $G(\gamma)$ denotes the left side of (2.9) and $F(\gamma)$ is given by equation (2.7). For the unique solvability of (2.8), the curves

$$Y = F(\gamma) \quad \text{and} \quad Y = G(\gamma)$$

must be as shown in Figure 2a and not as in Figure 2b or Figure 2c. The condition of Figure 2a is guaranteed by

$$G'(0) > \frac{4}{\sqrt{\pi}} .$$

A straightforward calculation shows

$$G'(0) = \frac{2}{Q}\sqrt{\frac{D}{t^*}}\,(\theta_s - \theta_r)$$

which will exceed $\frac{4}{\sqrt{\pi}}$ if t^* is sufficiently small. This means that if α and β are to be determined from a measurement of capillary pressure head, then *the surface measurement of h must be taken at a sufficiently small time in the flow experiment.* If the functions $F(\gamma)$ and $G(\gamma)$ are as in Figure 2c, we can see that, in addition two having two solutions, each of these two solutions may be very sensitive to small changes in the problem data.

(a)

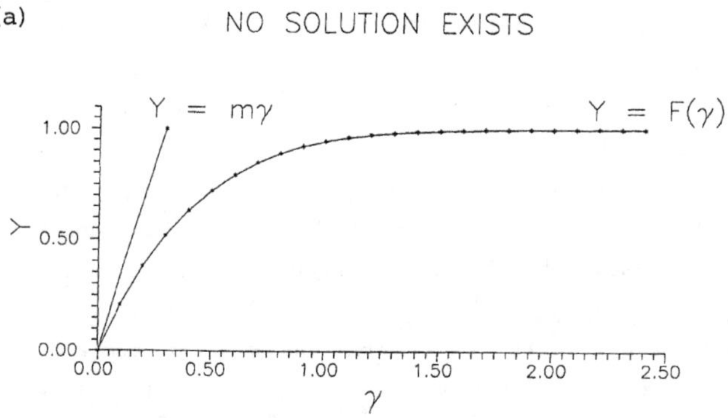

(b)

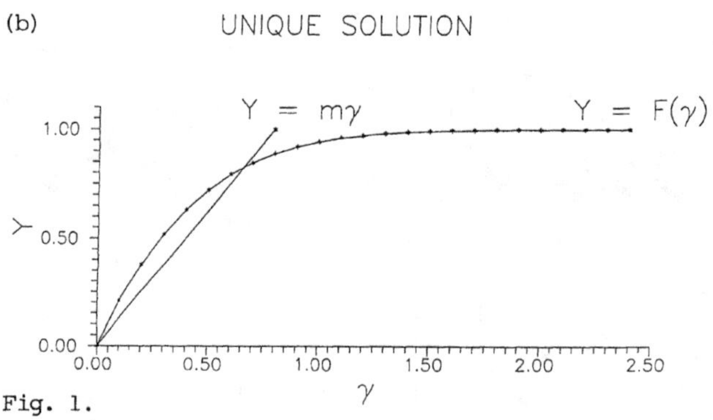

Fig. 1.

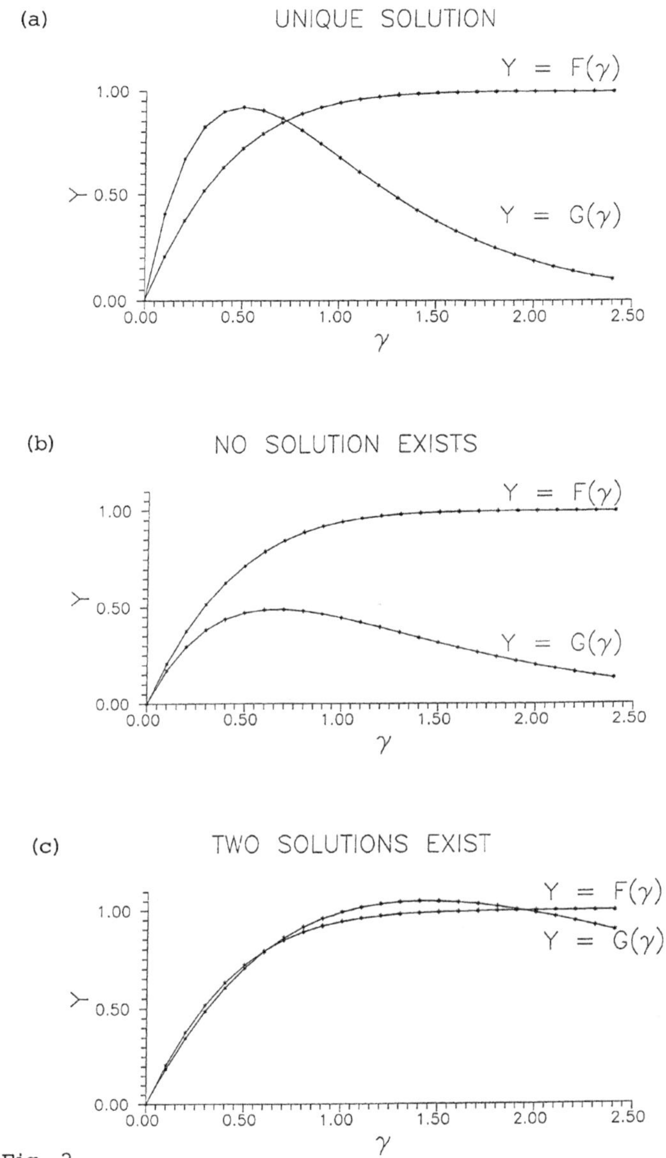
(a)
UNIQUE SOLUTION
Y = F(γ)
Y = G(γ)
1.00
Y 0.50
0.00
0.00 0.50 1.00 1.50 2.00 2.50
γ
(b)
NO SOLUTION EXISTS
Y = F(γ)
Y = G(γ)
1.00
Y 0.50
0.00
0.00 0.50 1.00 1.50 2.00 2.50
γ
(c)
TWO SOLUTIONS EXIST
Y = F(γ)
Y = G(γ)
1.00
Y 0.50
0.00
0.00 0.50 1.00 1.50 2.00 2.50
γ

Fig. 2.

References

Cannon, J.R. and D.W. Zachamnn Parameter determination in parabolic partial differential equations from over-specified boundary data. International Journal of Engineering Science. 20(6), 779-788, 1982.

Clothier, B.E. and I. White. Water diffusivity of a field soil. Soil Sci. Soc. Am. 46, 155-158, 1982.

Hornung,U. Identification of nonlinear soil physical parameters from an input-output experiment. In: P.Deuflhardt and E Harier(Editors), Workshop on Numerical Treatment of Inverse Problems in Differential and Integral Equations. Birkhauser, Boston, Mass, pp. 227-237, 1983.

Kool,J.B., J.C. Parker and M.Th.van Genuchten. Determining soil hydraulic properties from one-step outflow experiments by parameter estimation. I. Theory and numerical studies. Soil Sci. Soc. Am. J., 49, 1348-1354, 1985

Lomen,D.O. and A.W. Warrick. Linearized moisture flow with loss at the soil surface. Soil Sci. Soc. Amer. J. 42:396-400, 1974.

Milly, P.C.D. Estimation of Brooks-Corey parameters from water retention data. Water Resour. Res., 23(6), 1085-1089, 1987

Perroux, K.M., P.A.C. Raats and D.E. Smiles. Wetting moisture characteristics curves derived from constant-rate infiltration into this soil samples. Soil Sci Soc. Am., 46, 231-234, 1982

Philip, J.R. Steady Infiltration from buried point sources and spherical cavitities. Water Resour. Res. 4:1039-1047, 1968.

Philip, J.R. The Theory Infiltration. Advances in Hydrosciences 5:215-296, 1969.

Philip, J.R. General theorem on steady infiltration from surface sources, with application to point and line sources. Soil Sci. Soc. Amer. Proc. 35:399-401, 1971.

Poulovassilis,A., M. Polychronides and P. Kerkides. Evaluation of various computational schemes in calculating unsaturated hydraulic conductivity. Agric. Water Management, 13, 317-327, 1988.

Raats, P.A.C. Steady infiltration from line sources and furrows. Soil Sci Soc. Amer. Proc. 35:709-714, 1970.

Towner G.D. Analyzing one-step outflow experiments to calculate soil-water diffusivities using Gardner's equation. J. Soil Sci., 33, 351-364, 1982.

Warrick, A.W. Time dependent linearized infiltration. Soil Sci. Soc. Amer. Proc. 38:383-386, 1974.

White, I. Measured and approximate flux-concentration for absorption of water by soil. Soil Sci. Soc. Am. J., 43, 1074-1080, 1979.

White, I, D.E. Smiles and K.M. Perroux. Absorption of water by soil: The constant flux boundary condtion.

Zachmann, D.W. A mathematical treatment of infiltration from a line source into an inclined porous medium. Soil Sci. Soc. Amer. J. 42:685-688, 1978.

Zachmann, D.W. and A.W. Thomas. A mathematical investigation of steady infiltration from line sources. Soil Sci. Soc. Amer. Proc. 37:495-500, 1973.

Zachmann D.W., P.C.Duchateau and A. Klute. The calibration of the Richards flow equation for a draining colum by parameter identification. Soil Sci. Soc. Am. J. 45, 1012-1015, 1981.

Zachmann D.W., P.C.Duchateau and A. Klute. Simultaneous approximation of water capacity and soil hydraulic conductivity by parameter identification. Soil Sci. 134, 157-163, 1982.

GPSR Compliance
The European Union's (EU) General Product Safety Regulation (GPSR) is a set of rules that requires consumer products to be safe and our obligations to ensure this.

If you have any concerns about our products, you can contact us on

ProductSafety@springernature.com

In case Publisher is established outside the EU, the EU authorized representative is:

Springer Nature Customer Service Center GmbH
Europaplatz 3
69115 Heidelberg, Germany

www.ingramcontent.com/pod-product-compliance
Lightning Source LLC
LaVergne TN
LVHW011916200926
848484LV00002B/642
9780792316169